Contemporary's

Number Power

a real world approach to math

Robert Mitchell

CB

CONTEMPORARY BOOKS

a division of NTC/CONTEMPORARY PUBLISHING GROUP
Lincolnwood, Illinois USA

ISBN: 0-8092-2385-6

Published by Contemporary Books,
a division of NTC/Contemporary Publishing Group, Inc.,
4255 West Touhy Avenue,
Lincolnwood (Chicago), Illinois 60712-1975 U.S.A.
© 2000, 1991 by Robert Mitchell
5 6 7 8 9 10 11 12 13 14 15 16 ROV 14 13 12 11

Table of Contents

To the Student

Welcome to *Calculator Power:*

Calculator Power is designed to help you master the use of a hand-held calculator for all types of basic math problems. You will learn to use a calculator to add, to subtract, to multiply, and to divide. You'll discover how a calculator can simplify your work with whole numbers, decimal numbers, fractions, and percents. And you'll learn how a calculator can help you gain confidence in solving all kinds of word problems.

Calculator Power will help make using a calculator more enjoyable. Step-by-step examples will show how to use a calculator to solve math problems you're most likely to come across in daily life and on tests— whether educational tests or employment tests.

Several icons are used throughout *Calculator Power*. A calculator icon alerts you to exercises where the calculator is useful for problem solving. A consumer icon highlights pages where the calculator is important in your daily life as a consumer. A workplace icon indicates a focus on use of the calculator in workplace situations.

 calculator icon

 consumer icon

 workplace icon

An important skill for anyone who uses a calculator is to be on the lookout for keying errors, such as entering incorrect digits or accidentally pressing a digit twice. The most important skill for noticing keying errors is to anticipate about how large an answer should be. To develop that skill, *Calculator Power* emphasizes estimation and mental math throughout the book. The following icons will alert you to problems where using these skills will be especially helpful.

 estimation icon

 mental math icon

To get the most out of your work, do each exercise carefully. Check your answers with the answer key at the back of the book. Inside the back cover is a chart to help you keep track of your score on each exercise.

Calculator Pretest

This pretest will tell you which chapters of *Calculator Power* you need to concentrate on. Do all the problems that you can. There is no time limit. Check your answers with the answer key at the back of the book. Then use the chart at the end of the test to find pages where you need to work.

1. List the ten digit keys on a calculator.

For a four-function calculator, the key for the addition function is [+] and the key for the division function is [÷].

2. What is the key for the multiplication function?

3. What is the key for the subtraction function?

Match the following calculator keys and their meanings.

_____ 4. On/Clear all   a. [M+]

_____ 5. Add to memory   b. [ON/CA]

_____ 6. Clear memory   c. [MR]

_____ 7. Recall memory   d. [M–]

_____ 8. Subtract from memory   e. [MC]

9. Which order of key presses lets you show the value "3 dollars and 19 cents" as a decimal number in a calculator display?

   a. [3] [1] [9]

   b. [3] [·] [1] [9]

   c. [·] [3] [1] [9]

   d. [3] [1] [9] [·]

10. Which display shows the value "one hundred five thousand sixty?"

    a. 10500060.

    b. 100500060.

    c. 105060.

    d. 0.156

Write the value that a calculator displays for each list of key presses.

11. [7] [+] [5] [=]   12. [2] [×] [4] [=]

13. [12] [+] [2] [+] [1] [=]   14. [3] [×] [3] [×] [5] [=]

15. [15] [−] [10] [=]   16. [20] [÷] [10] [=]

17. Round 3,479 and 5,842 to the nearest hundred. Then add.

18. Round 443 and 79 to the nearest ten. Then subtract.

For problems 19–22, write each number as a decimal.

19. three-tenths

20. twenty-seven hundredths

21. one hundred five thousandths

22. three and two-thousandths

23. Round 14.2935 to the nearest thousandth.

24. Round 14.2935 to the nearest tenth.

25. Which number is greater, 4.3125 or $4\frac{3}{5}$?

26. Which number is less, 14.324 or $14\frac{5}{9}$?

27. What is 25% of 100?

28. The number 60 is what percent of 100?

29. The number 15 is 15% of what number?

Find the value of each expression.

30. $(3 + 4) \times 5 =$

31. $4 + \frac{6}{2} =$

32. $3 + (4 \times 5) =$

33. $25 + (15 \times 7) =$

34. $\dfrac{4 + 6}{2} =$

35. $(72 \div 9) \times 68 =$

Calculator Pretest Chart

This pretest can be used to identify calculator skills in which you are already proficient. If you miss only 1 question in a section of this test, you may not need further study in that chapter. However, to effectively master use of the calculator, we recommend that you work through the entire book. As you do, focus on the skills for the items that you missed.

PROBLEM NUMBER	SKILL AREA	PRACTICE PAGES
1, 2, 3, 4, 5, 6, 7, 8	calculator basics	7–10, 89–90
9, 10	displaying numbers	11–12
11, 13	adding	19–20
15	subtracting	21–22
12, 14	multiplying	31–32
16	dividing	33–34
17, 18	estimating whole numbers	27–28
19, 20, 21, 22	decimals	45–46
23, 24	rounding decimals	47–48
25, 26	comparing fractions and decimals	66–67
27	finding part of a whole	75–76
28	finding a percent	77–78
29	finding the whole	79–80
30, 31, 32, 33, 34, 35	evaluating arithmetic expressions	91–94

Building
Number
Power

BECOMING FAMILIAR WITH CALCULATORS

To **calculate** is to work with numbers. You calculate each time you add, subtract, multiply, or divide. You also calculate each time you estimate, perform mental math, or write arithmetic problems using paper and pencil.

In this section, you will learn the basics of using a calculator. By the end of this book, you will see that the best way to calculate combines estimation, mental math, and calculator use.

Calculator Basics

A **calculator** is an electronic device that makes it easy to work with numbers. When used carefully, a calculator is amazingly quick and accurate. The calculator pictured below may be similar to one you own or use.

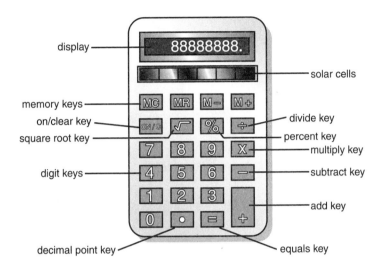

This calculator is an example of a **four-function, four-key memory calculator.** The four **function keys** are $+$, $-$, $\times$, and $\div$. The **memory keys** are $M+$, $M-$, MR, and MC.

You'll soon learn about all the keys shown above. For now, notice the group of ten **digit keys:** 0, 1, 2, 3, 4, 5, 6, 7, 8, and 9. Digit keys let you enter numbers on a calculator.

Make a drawing of your own calculator and label all the keys. You can make your own drawing or use one of the drawings below.

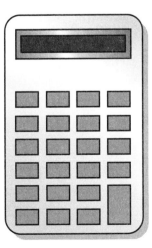

Battery Powered

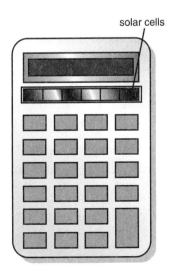

Solar Powered

Types of Calculators

A **four-function** calculator helps you solve problems involving addition, subtraction, multiplication, and division. Other types of calculators, used in banks, businesses, and laboratories, include business calculators, graphing calculators, and scientific calculators.

Solar-Powered Calculators

A **solar-powered** calculator contains a row of solar cells. These cells change light into electricity to make the calculator work. You can distinguish the solar cells from the display (where the numbers appear) by the cells' darker color.

Don't expect to see numbers appear on the solar cells! The cells' only purpose is to change light into electricity. A solar-powered calculator won't work in a dimly lit room.

Most solar-powered calculators have an ON key and an OFF key. Others simply have a cover that opens and closes, using light and darkness to turn the calculator on and off.

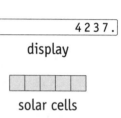

4237.

display

solar cells

Battery-Powered Calculators

A **battery-powered** calculator contains a battery— usually placed in the back of the calculator.

All battery-powered calculators have an ON key and an OFF key. On some calculators, these two functions are performed by a single ON/OFF key.

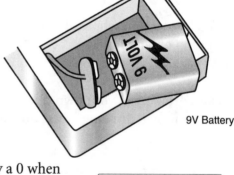

9V Battery

The Calculator Display

Most calculators display zero and a decimal point when they are turned on. If your calculator does not display a 0 when you press the ON key, your calculator has a problem.

- A solar-powered calculator may have a problem with its solar cells. You should return this calculator to the store.

- A battery-powered calculator may have a dead battery. Or someone may have forgotten to put a battery in!

The maximum number of digits a calculator can work with is determined by the size of its display. Most calculators have an 8-digit display.

0.

An "ON" display

24857394.

A displayed 8-digit number

Clear Keys

To start a new calculation or to change the number shown in the display, you use special erasing keys called **clear keys.** Different calculators use different clear key symbols. However, most calculators use one or two of the keys described below.

Learn the meaning of the clear keys on your calculator.

Key	Meaning	Function
CA	Clear All	The Clear All key erases the display. It also erases all parts of a calculation stored in the calculator and erases all values stored in the memory.
ON/C	On/Clear	Pressing ON/C turns the calculator on and erases the display.
CE	Clear Entry	The Clear Entry key erases the display only. It does not erase any calculations stored in the calculation or in memory.
CE/C	Clear Entry/Clear	Pressing CE/C once clears the display. Pressing CE/C twice clears the display, the memory, and other parts of a calculation stored in the calculator.
MC	Memory Clear	MC erases values stored in memory. It does not affect the display or parts of a calculation stored in the calculator. All the memory keys are explored in the chapter called Using the Calculator's Memory.

Discovery: On most calculators, you can also clear the display by turning the calculator off and then on again. Though it is not recommended, many people use this technique.

 Answer each of the following questions about your calculator.

1. Is your calculator solar powered or battery powered?

2. Which key do you press (if any) to turn your calculator on?

3. When turned on, what appears on your calculator's display?

4. After turning your calculator on, press the whole number keys in order: 1, 2, 3, 4, 5, 6, 7, 8, and 9. How many digits appear on your calculator display?

5. What is the largest number you can display on your calculator?

6. Which key do you press to erase your calculator's display?

The Decimal Point Key

The 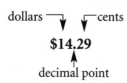 key is the **decimal point** key. One use of the decimal point key is to enter an amount of money. For example, to enter 14 dollars and 29 cents, you use the decimal point to separate dollars from cents. Cents always appear as 2 digits to the right of the decimal point.

dollars ⟍ ⟋cents

$14.29

↑
decimal point

Here are some points to remember.

- No dollar sign is displayed on a calculator.
  The amount $14.29 is entered as ⌷1⌷ ⌷4⌷ ⌷·⌷ ⌷2⌷ ⌷9⌷ and is
  displayed as

  | 14.29 |

- When you enter cents only, the calculator displays a 0 to the left of the decimal point.
  The amount $0.18 is entered as ⌷·⌷ ⌷1⌷ ⌷8⌷ and is displayed as

  | 0.18 |

- When you enter an amount between 1¢ and 9¢, put a 0 between the decimal point and the cents digit. As
  examples, you enter 4¢ as ⌷·⌷ ⌷0⌷ ⌷4⌷
  and you enter $3.08 as ⌷3⌷ ⌷·⌷ ⌷0⌷ ⌷8⌷.

  | 0.04 |
  | 3.08 |

 Enter each of the following money amounts on your calculator. Then, below each amount, write how that value appears on the calculator display. (Clear the display before each entry.)

7. $2.57 $17.81 $20.56 $0.49 $0.08 $0.10
   2.5 7

8. four dollars and nineteen cents   six dollars and twelve cents

9. twenty-seven cents   forty-two cents

10. one cent   eight cents

11. one dollar and six cents   two dollars and seven cents

More About Displayed Numbers

Think for a moment about things you've discovered about a calculator.

- Digit keys are used to enter numbers.

- You enter a number one digit at a time, starting with the left-hand digit.

- You can set the display to "0" by pressing a clear key.

- You press the decimal point key $\boxed{\cdot}$ to separate dollars from cents.

Here's another interesting fact: Calculators do not have a comma (,) key.

For example, to enter 1,960, you press $\boxed{1}$ $\boxed{9}$ $\boxed{6}$ $\boxed{0}$. *You do not enter a comma.*

EXAMPLE 1   To enter 82,500 on your calculator, press keys as shown below.

Press Keys	Display Reads
$\boxed{8}$	8.
$\boxed{2}$	82.
$\boxed{5}$	825.
$\boxed{0}$	8250.
$\boxed{0}$	82500.

Discovery
- Calculators display a decimal point to the right of a whole number.
- Many calculators do not display commas.
- A calculator may omit zeros at the right of a decimal number. The display 1367.5 means $1367.50.

EXAMPLE 2   To enter $1,367.50 on your calculator, press these keys.

	Press Keys	Display Reads
no comma is entered →	$\boxed{1}$	1.
	$\boxed{3}$	13.
	$\boxed{6}$	136.
	$\boxed{7}$	1367.
enter decimal point to separate dollars and cents →	$\boxed{\cdot}$	1367.
	$\boxed{5}$	1367.5
	$\boxed{0}$	1367.50

Rewrite each of the numbers. If the number is greater than 999, write it with a comma. The first is completed as an example.

no decimal point

1. [2 4 5 0 .]   2,450 ⟵   2. [8 7 5 .] _____

3. [4 0 5 6 .] _____   4. [3 9 4 5 0 .] _____

5. [1 8 3 2 .] _____   6. [2 9 6 0 9 .] _____

Write these money amounts with a dollar sign and comma if the amount is $1,000 or more.

7. [1 8 . 3 2] $18.32   8. [2 4 7 1 . 6 0] _____

9. [6 4 9 . 0 9] _____   10. [5 0 3 8 . 1 8] _____

Choose how each of the displayed numbers and money amounts would be written.

11. [3 4 0 .]   12. [9 0 0 .]

   a. thirty-four   a. nine
   b. three hundred forty   b. ninety
   c. three thousand, four hundred   c. nine hundred

13. [3 . 0 7]   14. [0 . 2 0]

   a. thirty-seven cents   a. twenty cents
   b. three dollars and three cents   b. two dollars
   c. three dollars and seven cents   c. twenty dollars

Enter each of the following numbers on your calculator. Write how each displayed number looks. Be sure to write the decimal point in the display.

15. ninety-five   [9 5 .]

16. two hundred forty-three   [   ]

17. three thousand, five hundred twenty-nine   [   ]

18. eight thousand, four hundred six   [   ]

19. fifteen dollars and eighty-two cents   [   ]

20. two hundred four dollars and nine cents   [   ]

Making the Display Speak!

Here is a creative game that can be played on a calculator. It might be an interesting introduction to calculators for you or someone you know.

1. On your calculator, see if you can discover what letter of the alphabet each digit most closely resembles when viewed upside down. Turn your calculator upside down, punch in the numbers, and fill in the letters you see.

Calculator digit	0	1	2	3	4	5	6	7	8
Letter resembled by upside-down digit	__	__	__	E	__	__	__	__	__

2. In the English alphabet, the vowels are *a, e, i, o, u,* and sometimes *y.* What are the three vowels in the upside-down calculator alphabet?

3. What word can be read on an upside-down display when each of the following numbers is entered?

   14 _____   345 _____   710 _____

4. Make a list of four two-letter words that can be written with "calculator letters." (**Hint:** Each word contains one vowel.)

5. Make a list of five three-letter words that can be written with "calculator letters."

WHOLE NUMBERS AND MONEY

Much of the math around us involves whole numbers and money. Calculators can help us work more effectively with large numbers and with groups of whole numbers. Calculators can also help us focus our efforts on thinking about how to solve problems.

In this part of *Calculator Power,* besides the digit keys, you will use the decimal point key ·, the clear (correct) key CE or ON/C, and the +, −, ×, ÷, and = keys to solve problems.

The first step in using a calculator for any problem is to press the clear key CE or ON/C. This clears the display and allows you to enter a new number. Be sure to clear your calculator's display as your first step in working all the problems in this book.

Adding Two Numbers

The Add Key and the Equals Key

The ⊞ key is called the **add key** and is used to add numbers. The ⊟ key is called the **equals key** and, when pressed, tells a calculator to display the answer to a calculation.

To see how the ⊞ key and the ⊟ key work together in a calculation, start by clearing the display. Then press the following keys in order:

[5] [+] [3] [=]

What answer does your calculator display show?

Adding Two Numbers

If your calculator displayed 8, you already know how to add two numbers! (**Remember:** Always clear your calculator display before beginning each new problem.)

EXAMPLE 1 To add 37 + 28 on your calculator, press the keys as shown.

Press Keys	Display Reads
[3] [7]	37.
[+]	37.
[2] [8]	28.
[=]	65.

> **Discovery**
> - Most displays do not show a + sign.
> - The answer appears only after you press =.
> - You must clear your calculator before you start a new problem.

ANSWER: 65

EXAMPLE 2 To add $2.38 and $1.09 on your calculator, press the keys as shown.

Press Keys	Display Reads
[2] [·] [3] [8]	2.38
[+]	2.38
[1] [·] [0] [9]	1.09
[=]	3.47

ANSWER: $3.47

Fill in the numbers and operations to show what keys you would press to solve each problem. Do not write the solution.

1. $8 + 4 =$ [8] [+] [4] [=]

2. $39 + 17 =$ ☐ ☐ ☐ ☐ ☐ ☐

3. $2{,}063 + 989 =$ ☐ ☐ ☐ ☐ ☐ ☐ ☐ ☐

4. $\$5.09 + \$4.26 =$ ☐ ☐ ☐ ☐ ☐ ☐ ☐ ☐ ☐

5. $\begin{array}{r} 128 \\ +\ 89 \\ \hline \end{array}$ ☐ ☐ ☐ ☐ ☐ ☐

6. $\begin{array}{r} \$7.90 \\ +\ 3.78 \\ \hline \end{array}$ ☐ ☐ ☐ ☐ ☐ ☐ ☐ ☐ ☐

7. thirty-five plus fourteen

   ☐ ☐ ☐ ☐ ☐ ☐

8. the sum of one hundred fifty-seven and sixty-one

   ☐ ☐ ☐ ☐ ☐ ☐ ☐

9. five thousand, two hundred nine plus two thousand, four hundred

   ☐ ☐ ☐ ☐ ☐ ☐ ☐ ☐ ☐

10. four dollars and fifty cents added to six dollars and twelve cents

    ☐ ☐ ☐ ☐ ☐ ☐ ☐ ☐ ☐

 Use your calculator to solve each of the following problems.

11. $41 + 18 =$ $450 + 207 =$ $1{,}950 + 872 =$

12. $\$0.98 + \$0.67 =$ $\$34.09 + \$20.98 =$ $\$245.75 + \$169.99 =$

13. What is the combined weight of the pickup and trailer?

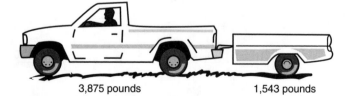

3,875 pounds   1,543 pounds

Keying Errors

Pressing the wrong key is called a **keying error.** When people use calculators, keying errors are very common. In fact, keying errors are so common that you must always be alert to them!

There are three main types of keying errors.

- pressing the wrong key
  Example: Pressing 8 2 instead of 9 2 .

- double keying—accidentally pressing the same key twice
  Example: Pressing 4 7 7 instead of 4 7 .

- transposing digits—pressing keys in the wrong order
  Example: Pressing 3 5 instead of 5 3 .

As you practice your calculator skills throughout this book, be careful to check your display so you avoid each type of keying error.

For each problem below, indicate by a check (✓) which type of error was made. Then fill in the blank keys to show the correct way to key each problem.

Problem to Solve	Keys Pressed	Wrong Key	Double Keying	Transposing Digits
1. 27 + 18 =	2 7 + 1 1 8 =	_____	_____	_____
Correct:	▢▢▢▢▢▢			
2. $20 + $39 =	2 0 + 9 3 =	_____	_____	_____
Correct:	▢▢▢▢▢			
3. 153 + 89 =	1 4 3 + 5 9 =	_____	_____	_____
Correct:	▢▢▢▢▢▢			
4. $0.67 + $0.49 =	· 7 6 + · 4 9 =	_____	_____	_____
Correct:	▢▢▢▢▢▢▢			

Note: When you have a number such as $20, as in problem 2 above, you can enter 20 or 20.00 on your calculator. If you enter 20.00, most calculators will display 20. as soon as you press the + key. When you have a number such as $0.67, as in problem 4 above, the calculator will display 0.67 whether you enter .67 or 0.67.

Deciding When to Use a Calculator

Try the experiment below. There are 24 problems, divided into three types. Try doing each problem three different ways:

- in your head
- with paper and pencil
- with a calculator

When you've finished all 24 problems, list the problems that were easiest for you to do in your head, using pencil and paper, or using a calculator.

Type A: One-Digit Addition

1. $6 + 0 =$ 2. $9 + 4 =$ 3. $4 + 5 =$ 4. $3 + 8 =$

5. $3 + 4 =$ 6. $7 + 6 =$ 7. $2 + 6 =$ 8. $8 + 9 =$

Type B: Two-Digit Addition

9. $30 + 20 =$ 10. $52 + 39 =$ 11. $51 + 46 =$ 12. $46 + 28 =$

13. $75 + 24 =$ 14. $57 + 49 =$ 15. $72 + 16 =$ 16. $86 + 47 =$

Type C: Three-Digit Addition

17. $\begin{array}{r} 348 \\ + 121 \\ \hline \end{array}$ 18. $\begin{array}{r} 658 \\ + 469 \\ \hline \end{array}$ 19. $\begin{array}{r} 635 \\ + 122 \\ \hline \end{array}$ 20. $\begin{array}{r} 801 \\ + 305 \\ \hline \end{array}$

21. $\begin{array}{r} 500 \\ + 300 \\ \hline \end{array}$ 22. $\begin{array}{r} \$886 \\ + \ \ 527 \\ \hline \end{array}$ 23. $\begin{array}{r} \$4.23 \\ + \ 2.45 \\ \hline \end{array}$ 24. $\begin{array}{r} \$6.93 \\ + \ 4.58 \\ \hline \end{array}$

25. In your head: _____

    Pencil and paper: _____

    Calculator: _____

26. Why do you think some problems are more easily done *without* the use of a calculator? Write your reasons below.

Adding Three or More Numbers

To add three or more numbers, a calculator adds the first two. Then it keeps adding the next number to the previous sum.

EXAMPLE 1   Add $9 + 8 + 6$.

Press Keys	Display Reads	
9	9.	
+	9.	
8	8.	
+	17.	← subtotal of $9 + 8$
6	6.	
=	23.	← total of $17 + 6$

ANSWER: 23

Discovery: Most calculators display a value each time the + key is pressed. That value is the result of the calculation(s) up to that moment.

Does your calculator display up-to-the-moment calculations? To find out, try Example 1 on your calculator.

EXAMPLE 2   Add.

$24.87
12.99
+ 8.35

Press Keys	Display Reads	
2 4 . 8 7	24.87	
+	24.87	
1 2 . 9 9	12.99	
+	37.86	← 24.87 + 12.99
8 . 3 5	8.35	
=	46.21	← 37.86 + 8.35

ANSWER: $46.21

For problems 1–3, use the $\boxed{=}$ key only once when you fill in the key symbols for the problem.

1. $12 + 9 + 8 =$

$\boxed{1}\ \boxed{2}\ \boxed{+}\ \boxed{9}\ \boxed{+}\ \boxed{8}\ \boxed{=}$

2. $35 + 27 + 9 =$

$\boxed{}\ \boxed{}\ \boxed{}\ \boxed{}\ \boxed{}\ \boxed{}\ \boxed{}$

3. $\$1.53 + \$0.94 + \$0.58 =$

$\boxed{}\ \boxed{}\ \boxed{}\ \boxed{}\ \boxed{}\ \boxed{}\ \boxed{}$
$\boxed{}\ \boxed{}\ \boxed{}\ \boxed{}\ \boxed{}$

Estimating: Spotting Keying Errors Using Rounded Numbers

Suppose you use your calculator and find the sum at the right. A quick check to identify keying errors is to round each number and use mental math to add the rounded values.

$$\begin{array}{r} 59 \\ 81 \\ +\ 37 \\ \hline 147 \end{array}$$

	rounds to	
59	$\rightarrow$	60
81	$\rightarrow$	80
+ 37	$\rightarrow$	+ 40
	Estimate is:	180

The estimated value 180 is not close to 147, so use your calculator and try the problem again.

$$59 + 81 + 37 = 177$$

The value 177 agrees with the estimate of 180. The value 147 was incorrect, probably due to a keying error.

....

A student used a calculator to solve the following problems. Because of keying errors, four of the problems have *wrong answers*. Use estimation to spot the four problems with wrong answers. Then use your calculator to find correct answers.

4.	5.	6.	7.
$\begin{array}{r} 28 \\ 17 \\ +\ 9 \\ \hline 54 \end{array}$	$\begin{array}{r} 354 \\ 227 \\ +100 \\ \hline 681 \end{array}$	$\begin{array}{r} \$8.43 \\ 3.57 \\ +\ \ 0.98 \\ \hline \$15.98 \end{array}$	$\begin{array}{r} \$365 \\ 154 \\ +\ \ 116 \\ \hline \$935 \end{array}$

8.	9.	10.	11.
$\begin{array}{r} 231 \\ 106 \\ +\ \ 53 \\ \hline 290 \end{array}$	$\begin{array}{r} \$28.90 \\ 17.83 \\ +\ \ 13.58 \\ \hline \$60.31 \end{array}$	$\begin{array}{r} \$6.90 \\ 8.23 \\ +\ \ 4.50 \\ \hline \$19.63 \end{array}$	$\begin{array}{r} 5,675 \\ 1,476 \\ +\ \ 850 \\ \hline 2,794 \end{array}$

Subtracting Two or More Numbers

The $\boxed{-}$ key is called the **subtract key** and is used to subtract one number from another.

EXAMPLE 1 To subtract 37 from 108 on your calculator, press keys as shown.

Press Keys	Display Reads
$\boxed{1}\boxed{0}\boxed{8}$	108.
$\boxed{-}$	108.
$\boxed{3}\boxed{7}$	37.
$\boxed{=}$	71.

ANSWER: 71

You can use your calculator for an expression such as $20 - 3 - 5 - 3$ by pressing $\boxed{-}$ before each new subtraction. Press $\boxed{=}$ only once, after you enter the last number.

EXAMPLE 2 Stan paid $11.98 for a hammer plus $0.60 tax. How much change did Stan get from a $20 bill?

To solve, start with $20 and subtract each amount.

Press Keys	Display Reads
$\boxed{2}\boxed{0}\boxed{\cdot}\boxed{0}\boxed{0}$	20.00
$\boxed{-}$	20.
$\boxed{1}\boxed{1}\boxed{\cdot}\boxed{9}\boxed{8}$	11.98
$\boxed{-}$	8.02
$\boxed{\cdot}\boxed{6}\boxed{0}$	0.60
$\boxed{=}$	7.42

ANSWER: $7.42

Reminder: When you have numbers such as $20, you can enter 20.00 or just 20. Many people prefer to enter all of the 0s as a reminder that they are working with dollars and cents.

Discovery:
Try the following subtraction problem on your calculator:

You subtract	The display reads
$0.35 – $0.15	0.2

The answer is $0.20, but the display reads 0.2.
- A calculator does not display a 0 that is at the right-hand end of the decimal part of an answer.
- When you write an answer in money, remember to write two digits in the decimal part.

Fill in the numbers and operations to show what keys you would press to solve each problem. Do not write the solution.

1. the difference between eighty-six and forty-nine

   ☐ ☐ ☐ ☐ ☐ ☐

2. two hundred seventeen subtracted from five hundred eight

   ☐ ☐ ☐ ☐ ☐ ☐ ☐ ☐

3. ten dollars minus seven dollars and ninety-two cents

   ☐ ☐ ☐ ☐ ☐ ☐ ☐ ☐ ☐ ☐

4. eight dollars minus three dollars and eleven cents

   ☐ ☐ ☐ ☐ ☐ ☐ ☐ ☐ ☐

5. Show another set of numbers and operations for problems 4 and 5.

Keying errors led one student to three *incorrect* answers in the problems below. See if you can find them. Then use your calculator to find correct answers.

6.	7.	8.	9.	10.
27	107	422	$20.00	$124.00
− 9	− 79	− 106	− 12.49	− 82.46
36	28	88	$7.51	$206.46

Use your calculator to solve each of the following problems.

11. $47 - 23 - 9 =$ $119 - 57 - 28 =$ $231 - 103 - 37 =$

12. $\$20.00 - \$13.89 - \$3.19 =$ $\$25.00 - \$9.08 - \$4.52 =$

13. As pictured at the right, how many pounds heavier than the sports car is the van?

4,138 pounds

2,869 pounds

14. Mari paid for her lunch, shown at the right, with a 10-dollar bill. If her tax is $0.35, how much change should Mari be given?

Coffee
$1.25

Ham & Cheese Sandwich
$4.45

Calculators and Word Problems

Have you ever thought, "Wow, that calculator is really fast!" Yet, as fast and amazing as it is, a calculator can't tell you

- which numbers are important
- whether to add or to subtract
- whether the answer you compute is correct

Five Steps for Solving Word Problems

When solving word problems, you may find it useful to follow this problem-solving approach.

STEP 1   Read the word problem and find the key information.

STEP 2   Estimate an answer.

STEP 3   Choose the operation (add, subtract, multiply, or divide) and set up a calculation.

STEP 4   Perform and check the calculation(s).

STEP 5   Reread the word problem and see if your answer makes sense.

Nathan traded in his Ford for a new Honda. Although the sticker price was $23,450, the salesperson lowered this price by $1,875. Nathan was also given a trade-in allowance of $7,250, although the Ford was only worth $6,525. Using his trade-in as a down payment, Nathan wanted to know how much he still owed for the Honda.

 Decide whether "mental skills" or "calculator skills" are most important in helping you answer each question below.

1. What is the key information?   mental math   calculator

2. What rounded numbers should you use to estimate an answer?   mental math   calculator

3. What operations should you use?   mental math   calculator

4. What answer results from each operation you perform?   mental math   calculator

5. Does your answer make sense?   mental math   calculator

Building Confidence with Word Problems

Carefully read word problems A and B and do the exercises that follow.

A. Last weekend, Hamburger Heaven sold $1,843 worth of hamburgers, $1,257 worth of fries, and $1,385 worth of drinks. How much money did Hamburger Heaven take in on these three products last weekend?

B. Last weekend, Hamburger Heaven sold $50 worth of hamburgers, $20 worth of fries, and $30 worth of drinks. How much money did Hamburger Heaven take in on these three products last weekend?

For problems 6 and 7, check (✓) the one answer that most closely expresses your opinion.

6. Which problem above, A or B, seems more difficult?

   A _____   B _____   About the same _____

7. If you checked A or B above, why do you think that problem was more difficult than the other?

   _____ Different situations are described.

   _____ The harder problem contains larger numbers.

   _____ The harder problem is longer.

   If you're like most students, you think problem A is more difficult. Why? Most likely because it contains larger numbers.

   This is a natural conclusion. After all, we usually think that larger numbers are harder to work with than smaller numbers.

   By using a calculator, you can concentrate on solving the problem. All types of numbers—large and small—become easier to work with.

   Here's how to practice solving word problems using a calculator.

   - Read the problem for understanding.
   - Concentrate on problem-solving steps 1, 2, 3, and 5 listed on the previous page.
   - Have confidence that you and your calculator can do any calculation correctly.

8. In your own words, tell why you think a calculator can help you become a better word-problem solver.

The Importance of Estimating

Herald Gazette

Read All About It!
Math Common Sense Rescues Shopper
Lilly Sharp was sharper than a cash register yesterday when she was about to be charged $45.29 for some purchases. Saved by her common sense, Lilly insisted that the cashier ring up her total again. This time the total was $26.29. Sharp reported, "I bought things for $11.59, $9.95, and $4.75. I knew I spent less than 12 dollars plus 10 dollars plus 5 dollars. I can't afford not to be careful!"

Lilly may be fictional, but her story illustrates an important fact.

- Being able to spot an incorrect answer is a very valuable skill.

Sometimes you can spot a wrong answer by using **math intuition** or **common sense.** You can **estimate** or **approximate** a value to check a calculator or cash register result.

- Math intuition comes from thinking about what an answer should be. Estimation told Lilly that her purchases cost less than $27.

- To estimate, you use rounded numbers. Rounded numbers are easier to work with than exact values.

EXAMPLE: On another occasion, Lilly paid for a $28.63 purchase with a $50 bill. Is $12.37 a reasonable amount of change?

Lilly expected to get about $20 in change. Lilly thought, "$28.63 is about $30, and $50 − $30 is $20. My change should be about $20. So $12.37 is not enough change!"

As Lilly discovered, the best protection against calculator error is the ability to spot a wrong answer. And estimating is your most helpful tool.

Throughout the rest of this book, you'll use estimation to check calculator answers. By using estimation to catch keying errors, you'll learn to use a calculator accurately and confidently.

Estimating with Whole Numbers

To estimate, replace each number in a problem with a **rounded number.** A rounded number has zeros to the right of a chosen place value, such as the tens place, the hundreds place, the thousands place, and so on.

For example, $28.63 rounded to the nearest ten dollars is $30.00.

EXAMPLE Estimate an answer for the addition problem at the right.

To do this, round each number to the nearest hundred. Then add.

Problem	Estimate
912	900
786	800
+ 328	+ 300
2,026 ← close →	2,000

Estimate an answer to each problem by rounding. Round each number to the nearest ten. The first one is done for you.

		Estimate		Estimate		Estimate
1.	89	90	48		92	
	+72	+ 70	+ 33		− 41	
		160				

Round each number to the nearest hundred.

		Estimate		Estimate		Estimate
2.	488	500	897		$724	
	− 213	− 200	+108		− 336	
		300				

Round each number to the nearest thousand.

		Estimate		Estimate		Estimate
3.	4,927	5,000	7,849		12,193	
	+ 2,083	+2,000	+4,906		− 7,826	
		7,000				

Find the best estimate for the answer to each problem below.

Computation Problems

_____ 4. $38 + 54 =$

_____ 5. $92 - 33 =$

_____ 6. $22 + 19 + 9 =$

_____ 7. $189 - 93 =$

_____ 8. $234 + 512 =$

Estimates

a. 50

b. 100

c. 700

d. 90

e. 60

Estimation with Addition and Subtraction

Practicing with estimation will help you with calculator use. Estimating an answer will help you

- choose an operation when you're not sure what to do

- detect a keying error when you have calculated incorrectly

In each problem below, substitute rounded numbers and compute an estimate. Use your estimate as a clue to choose the correct answer from the choices given.

1. Joyce paid $10.00 for a hairbrush that cost $6.19, including tax. How much change should Joyce receive?

    Substitute $6.00 for $6.19.

    a. $1.19   **b.** $3.81   **c.** $5.79

2. Attendance figures for three basketball games were as follows: Thursday—3,894, Friday—2,179, and Saturday—1,946. How many tickets were sold for these three games?

    Substitute 4,000 for 3,894, substitute 2,000 for 2,179, and substitute 2,000 for 1,946.

    a. 6,849   **b.** 7,139   **c.** 8,019

3. Friday's dinner cost Jan $12.79 for lasagna, $4.75 for dessert, and $1.20 for coffee. Not including tax, what was the cost of Jan's meal?

    Substitute $13.00 for $12.79, substitute $5.00 for $4.75, and substitute $1.00 for $1.20.

    a. $18.74   **b.** $22.44   **c.** $24.84

For each of the following problems, use your calculator to compute an exact answer. Then estimate an answer in your head or with pencil and paper as you did on page 26. Did you make any keying errors? See how close your estimate is to the exact answer.

4. Count the total calories in the following meal:

ham and cheese sandwich:	397 calories
one glass of whole milk:	207 calories
one piece of cherry pie:	288 calories

    _____ exact

    _____ estimate

Problems 5 and 6 are based on the following diagram.

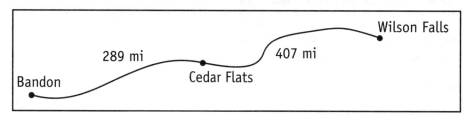

5. What is the distance between Bandon and Wilson Falls?

_____
exact

_____
estimate

6. How much farther is the distance from Cedar Flats to Wilson Falls than the distance from Cedar Flats to Bandon?

_____
exact

_____
estimate

Be careful. Problems 7–9 contain more information than you need.

7. At a yard sale on Sunday, Jesse bought tools for $23.88 and an electric motor for $19.95. He also bought a small vise. After writing a check for $60.00, Jesse received $3.89 in change. How much did Jesse spend in all? (**Hint:** Necessary information: $3.89, $60.00; Extra information: $23.88, $19.95)

_____
exact

_____
estimate

8. Of the 91 people entered in the Senior Jog-A-Thon, 19 are over 65 years of age and only 29 are under 50 years of age. More than half, 49 to be exact, are women. Given these figures, determine the number of men entered in this year's Senior Jog-A-Thon.

_____
exact

_____
estimate

9. On the invoice shown at the right, what is the total cost of the three newly purchased items, not counting the cost of the insurance, shipping, and C.O.D. service charges?

_____
exact

_____
estimate

Anderson Manufacturing Company

Item #	Description	Quantity	Amount
28474B	Brass Table Lamp	1	$89.96
10277C	Oval Wall Mirror	1	$38.75
03278E	4-Shelf Bookcase	1	$48.89
	Insurance		$4.75
	Shipping		$23.85
	C.O.D. Service Charge		$8.00
		TOTAL COST	$176.31

Application: Balancing a Checkbook

 Balancing a checkbook involves keeping a careful record of all the transactions: writing checks, making deposits, and paying bank service charges. A calculator can save you lots of time, especially if you need to recheck your work when your checkbook and bank statements don't match.

Recording Transactions

Below is a sample **check register,** a page in a checkbook. The account holder, Leona Jenson, is not charged a fee for the checks that she writes or for the deposits that she makes.

Number	Date	Description of Transaction	Payment/ Debit (−)	✓ T	Fee (−)	Deposit/ Credit (+)	Balance $1568|43
202	6/1	North Street Apartments	$ 825|00		$	$	743|43
203	6/4	Amy's Market	39|74				
204	6/8	Import Auto Repair	109|66				
	6/15	Payroll Deposit				1442|45	
205	6/19	Value Pharmacy	13|29				
206	6/21	Gazette Times	18|75				

1. Use your calculator to determine each daily balance in Leona's checking account. Record your answers in the BALANCE column.

   • Subtract each check (PAYMENT/DEBIT column) from the BALANCE column.

   • Add each deposit (DEPOSIT/CREDIT column) to the BALANCE column.

2. Record the following new transactions in the register. Then compute Leona's daily balance up through June 30 and enter it on each line in the BALANCE column.

   Check 207, written to Nelsen's on June 23 for $39.83
   Check 208, written to Hi-Ho Foods on June 24 for $63.79
   Check reorder charge of $26.50 on June 26
   Payroll deposit of $1442.40 on June 30

Reconciling a Bank Statement

At the end of each month, the bank sends each customer a **bank statement.** A bank statement shows the checking account balance at the beginning and at the end of each month. It also shows deposits and checks that come back to the bank. Special bank service charges are also shown.

To **reconcile a bank statement** is to make sure it agrees with information on the check register.

The example bank statement below refers to Leona Jenson's check register shown on page 29.

> **Discovery:** Many people like to check a calculation right after it's made. In a checkbook you can do this by
> - doing each addition or subtraction twice
> - checking addition with subtraction, and checking subtraction with addition

BANK STATEMENT

06/1–06/30 Leona Jenson Checking Account	Beginning Balance 1568.43	Deposits and Credits 2884.90	Withdrawals and Debits 1083.57	Ending Balance 3369.76

Monthly Service Charge		5.50
Check Reorder Charge	6/26	26.50

CHECKS

Nbr	Date	Amount	Nbr	Date	Amount
202	06/01	825.00	207*	06/23	39.83
204*	06/08	109.66	208	06/24	63.79
205	06/19	13.29			

*Gap in check sequence (The check before it has not been cashed.)

3. Follow this step-by-step method to reconcile Leona's bank statement on this page with her check register on page 29.

   STEP 1   For each check that's listed on the bank statement on this page, place a check (✓) in the ✓ column of the check register.

   STEP 2   Add the amounts of the two June checks that Leona wrote which are not listed on the bank statement.

   STEP 3   Subtract the sum found in Step 2 from the June 30 Ending Balance reported on the bank statement.

   STEP 4   Subtract the service charge (on the bank statement) from the June 30 balance (on the check register).

   STEP 5   Compare the amount computed in Step 3 to that computed in Step 4. Are they the same? If not, check your calculations for both the check register and this exercise.

Multiplying Two or More Numbers

The $\times$ key is called the **multiply key** and is used to multiply two numbers.

EXAMPLE 1 To multiply 235 times 54 on your calculator, press keys as shown.

Press Keys	Display Reads
2 3 5	235.
×	235.
5 4	54.
=	12690.

ANSWER: 12,690

As shown in Example 2 below, more than two numbers can be multiplied by pressing $\times$ before each new multiplication. As in addition and subtraction, you press $=$ only once, after entering the final number to be multiplied.

Reminder: Clear your calculator display before you start a new problem.

EXAMPLE 2 Oil is on sale for $2.19 per quart. How much do 7 cases of oil cost if each case contains 12 quarts?

To solve, multiply $2.19 by the total number of quarts—the number of quarts per case (12) times the number of cases (7).

$2.19 \times 12 \times 7 =$

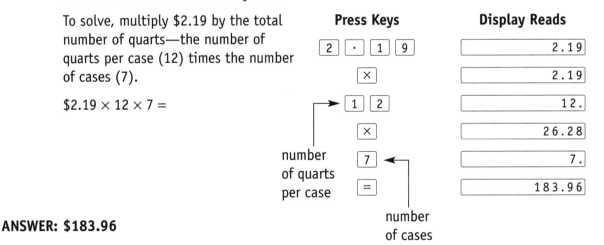

ANSWER: $183.96

Fill in the numbers and the symbols to show how you would key in each problem. Do not write the solution.

1. the product of seventy-six and forty

   ☐ ☐ ☒ ☐ ☐ ☐=

2. one hundred six times eighty-eight

   ☐ ☐ ☐ ☐ ☐ ☐ ☐

3. multiply nine dollars and four cents by seven

   ☐ ☐ ☐ ☐ ☐ ☐ ☐

 Compute both an exact answer and an estimate for each problem below. Remember that estimates can help you see if you have made keying errors.

- **Round numbers between 10 and 100 to the nearest ten.**

- **Round numbers larger than 100 to the nearest hundred.**

	Exact	Estimate	Exact	Estimate	Exact	Estimate
4.	48 × 32	50 × 30 1,500	67 × 59		88 × 19	

	Exact	Estimate	Exact	Estimate	Exact	Estimate
5.	192 × 57	200 × 60 12,000	289 × 32		207 × 74	

 Use your calculator to solve each of the following problems.

6. $7 \times 8 \times 5 =$ $19 \times 8 \times 6 =$ $23 \times 17 \times 5 =$

7. $74 \times 28 \times 3 =$ $143 \times 7 \times 2 =$ $258 \times 127 \times 3 =$

8. $\$23.45 \times 6 =$ $\$12.74 \times 20 =$ $\$52.09 \times 18 =$

9. At a Labor Day sale, how much will Brad pay for four cases of orange juice concentrate?

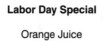

Labor Day Special

Orange Juice
Concentrate

$.89 per can

1 case

24 cans

Dividing One Number by Another

To divide is to see how many times one number (called the **divisor**) can go into a second number (called the **dividend**). The answer to a division problem is called the **quotient.**

As a quick review, the three ways that division problems are usually written are shown below.

$$27\overline{)675}^{\,25}$$

$$675 \div 27 = 25$$
$$\text{dividend} \div \text{divisor} = \text{quotient}$$

$$\frac{675}{27} = 25$$

$$\text{divisor}\overline{)\text{dividend}}^{\,\text{quotient}}$$

$$\frac{\text{dividend}}{\text{divisor}} = \text{quotient}$$

To divide on your calculator, follow these steps.

STEP 1 Enter the dividend.

STEP 2 Press ÷ , the divide key.

STEP 3 Enter the divisor.

STEP 4 Press = .

EXAMPLE 1 $49\overline{)833}$

To solve, press keys as shown.

Press Keys	Display Reads
8 3 3	8 3 3 .
÷	8 3 3 .
4 9	4 9 .
=	1 7 .

ANSWER: 17

EXAMPLE 2 $\$39.76 \div 7 =$

Clear your calculator display. Then press keys as shown.

Press Keys	Display Reads
3 9 · 7 6	39.76
÷	39.76
7	7 .
=	5.68

ANSWER: $5.68

EXAMPLE 3 $\dfrac{1023}{31}$

To solve, press keys as shown at the right.

Press Keys	Display Reads
1 0 2 3	1023.
÷	1023.
3 1	31.
=	33.

ANSWER: 33

....

Identify the dividend and divisor in each problem below. Then show how you would key in each problem. Do not write the answers.

1. 19)152   dividend _____
   divisor _____
   ☐ ☐ ☐ ☐ ☐ ☐

2. $19.68 ÷ 8   dividend _____
   divisor _____
   ☐ ☐ ☐ ☐ ☐ ☐ ☐

3. $\dfrac{361.56}{23}$ dividend _____
   divisor _____
   ☐ ☐ ☐ ☐ ☐ ☐ ☐ ☐

4. 27)576   dividend _____
   divisor _____
   ☐ ☐ ☐ ☐ ☐ ☐

  Use your calculator to compute an exact answer to each problem below. Then estimate an answer.

Exact
5. 418 ÷ 19 =

Exact
693 ÷ 21 =

Exact
288 ÷ 18 =

Estimate
400 ÷ 20 = 20

Estimate

Estimate

 **Use your calculator to solve each of the following problems.**

6. 11)495   14)364   19)$912   28)$1,708

7. Two hundred fifty-six people plan to attend this year's church picnic. If one table can seat eight people, how many tables will be needed in all?

8. Mrs. Owens gives 21 music lessons each week for a total of $735. How much does she charge for each lesson?

Using Math Common Sense with Multiplication and Division

Read the problems and use your math common sense to choose the correct answer. It's not necessary to do a computation.

1. If he takes home an average of $38 in tips each day, how many working days will it take Phil to save $570 from tips?

   a. multiplication: $21,660
   b. division: $15

2. There are 16 cups in one gallon. If a cup can hold 8 ounces of liquid, how many ounces can a gallon hold?

   a. multiplication: 128 oz
   b. division: 2 oz

3. When cherries are selling for $2.80 per pound, how much do 7 pounds of cherries cost?

   a. multiplication: $19.60
   b. division: $0.40

4. If Frank types at the rate of 55 words per minute, how many minutes will it take him to type a report containing 1,760 words?

   a. multiplication: 96,800 min
   b. division: 32 min

> **Discovery:** On some word problems, you may not know whether to multiply or divide. Do both on your calculator! Only one of the two answers will make sense.

Estimate answers for problems 5–8. Then use your calculator to compute the exact answers.

5. Last week Joni worked a total of 39 hours. If her hourly pay rate is $8.75, how much were Joni's earnings last week before taxes?

   _____ estimate
   _____ exact

6. The Oregon state lottery prize of $19,972,800 is to be divided equally among 9 winners. Determine each winner's share.

   _____ estimate
   _____ exact

7. West Wide Tool Manufacturing makes and ships 4,392 wrenches each day. If a full shipping box contains 18 wrenches, how many boxes are needed each day by West Side Tool?

_____
estimate

_____
exact

8. The professional wrestling match between The Slippery Savage and Pretty Boy MacDougal attracted 7,090 fans. Each paid $4.75 for a ticket. What amount of money was brought in from ticket sales for this match?

_____
estimate

_____
exact

 Problems 9–12 refer to the following story. Each question requires its own specific information. This information may be entirely in the story, or it may be partly in the question itself. Choose information carefully as you answer each question.

Maria works as a secretary at Jackson Bookkeeping Services. For her work she receives $330 each week. Only two years ago, though, Maria worked as a receptionist at an office and earned $268 each week.

Maria's husband, Mel, earns $34,580 each year as a lab technician at Lewis Electronic Fabricators.

9. When Maria worked as a receptionist, how much could she earn in one year if she worked all 52 weeks?

10. In her job as a secretary, how much does Maria now earn each year if she is paid for 52 weeks of work?

11. Assuming she works 40 hours each week, what is Maria paid hourly as a secretary?

12. Determine Mel's average weekly pay. Assume he is paid for 52 weeks.

Multistep Word Problems

Solving some problems involves more than one step.

EXAMPLE 1   Ellen is buying 6 bottles of hair conditioner on sale. Each bottle costs $2.89. How much change will Ellen receive if she pays the clerk with a $20 bill?

Solving this problem takes two steps: multiplication and subtraction.

STEP 1   Multiply to find the cost of the conditioner.

$2.89 \times 6 = 17.34

STEP 2   Subtract to find Ellen's change from $20.

$20.00 - $17.34 = 2.66

There are two methods for using a calculator to solve a multistep problem.

- **Method I:** Use pencil and paper to record and use the results from each step.

   STEP 1   Multiply to find the cost of the conditioner. Then write the cost, $17.34, on paper so you don't forget it.

   STEP 2   Clear the display and enter $20.00. Subtract $17.34 from $20.00 to get the answer $2.66.

- **Method II:** Use **memory keys** to save and use the results of each step. Pencil and paper are not needed. The discussion of memory keys begins on page 88.

 Use Method I to solve each of the following multistep problems.

1. At the Friday Halloween sale, Judy bought three hats on sale for $14.89 each. If she gave the clerk $50.00, how much change should Judy be given?

    > **Hint:**
    >
    > cost of hats = _____
    >
    > change = _____

2. Linda cooks at Mom's Breakfast restaurant. Last week she used 1,728 eggs. How many cases of eggs did she use if each case contains 24 cartons and each carton contains 12 eggs?

    > **Hint:**
    >
    > number of eggs in each case = _____
    >
    > number of cases used = _____

3. In the *News Tribune*, a weekend classified ad costs $15.75 for the first 16 words. Additional words are charged at the rate of $0.95 per word. At this rate, how much will Randi pay for a 24-word ad?

4. The brochure at the right lists the child-care charges at the Little Bunnies Center. Li leaves her daughter at the center three full days each week and two half days. What amount is Li charged each week for child care?

Little Bunnies Center	
Child-Care Rates	
Full Day	$28.75
Half Day	$17.75
Hourly	$5.50

Discovery: When adding a list of numbers, you may make a mistake on one entry. If you do, press the clear entry key [CE] to clear the display. Then reenter the number correctly and continue adding. In this way, **you do not need to redo the whole problem.**

5. Bea makes and sells flower baskets. Out of a total of 405 flowers, she plans to use 119 for large displays. The remaining flowers will be divided equally among 26 small flower baskets. How many flowers will be in each of these small flower baskets?

6. As assistant manager of Burger Supreme, Adah keeps track of weekly sales figures. The figures for the week of March 3 are shown at the right. By how much is this week's total below Burger Supreme's weekly average of $85,283.28?

Sales Figures Starting 3/3	
Monday	$11,472.25
Tuesday	$11,784.04
Wednesday	$11,800.34
Thursday	$12,028.50
Friday	$12,135.79
Saturday	$12,465.93
Sunday	$12,233.34
Total:	

Application: Depositing Money in a Checking Account

Retail businesses (those which sell directly to the public) take in money and checks every day. Accurately counting these receipts and depositing them in the company checking account is an important employee responsibility. Needless to say, a calculator can be a big help in a job like this.

The following definitions are used on a **checking account deposit slip.**
- **Currency:** paper money such as five-dollar bills
- **Coin:** pennies, nickels, dimes, and so on
- **Cash Received:** money that you request the bank to give back to you from your deposit
- **Net Deposit:** the amount you actually deposit, which is the TOTAL minus CASH RECEIVED

FOR DEPOSIT TO THE ACCOUNT OF		

SALLY'S ICE CREAMERY
4123 Loop Blvd.
Chicago, Illinois 60601

CASH	CURRENCY			**036**
	COIN			
LIST CHECKS SINGLY				24-7938/3239

DATE _____
DEPOSITS MAY NOT BE AVAILABLE FOR IMMEDIATE WITHDRAWAL

TOTAL FROM OTHER SIDE		USE OTHER SIDE FOR ADDITIONAL LISTING
TOTAL		
LESS CASH RECEIVED		
NET DEPOSIT		*BE SURE EACH ITEM IS PROPERLY ENDORSED*

_____
SIGN HERE FOR CASH RECEIVED

LOOP BANK

CHICAGO OFFICE
23 Jordan Street
Chicago, Illinois 60623

CHECKS AND OTHER ITEMS ARE RECEIVED SUBJECT TO THE PROVISIONS OF THE UNIFORM COMMERCIAL CODE OR ANY APPLICABLE COLLECTIONS AGREEMENT

Each Tuesday morning, Brett deposits the business receipts of Sally's Ice Creamery. Today's deposit consists of the money indicated below.

Coins	Currency	Checks	
150 pennies	1,168 one-dollar bills	#24-72	$44.89
80 nickels	137 five-dollar bills	#31-29	$39.85
250 dimes	29 ten-dollar bills	#42-16	$120.00
160 quarters	33 twenty-dollar bills		

Use your calculator and write each of the following on the deposit slip.

1. the total amount of CURRENCY
2. the total amount of COIN
3. the amount of each CHECK
4. the TOTAL (CURRENCY, COIN, and CHECKS). The amount of CASH RECEIVED is zero, so write the TOTAL in the space labeled NET DEPOSIT.

Calculator Division and Remainders

Remainders in Division

Many division problems contain a remainder as part of the answer. For example, if you divide 17 sheets of drawing paper among 5 students, each student gets 3 sheets. Since 5×3 is 15, 2 sheets are left over. Dividing 17 by 5 gives an answer of 3 with a remainder of 2.

There are three ways to write a remainder.

1. as a whole number following r

$$\begin{array}{r} 3\,r\,2 \\ 5\overline{)17} \\ \underline{15} \\ 2 \end{array}$$

2. as the numerator of a fraction

$$\begin{array}{r} 3\frac{2}{5} \\ 5\overline{)17} \\ \underline{15} \\ 2 \end{array}$$

3. as a decimal fraction

$$\begin{array}{r} 3.4 \\ 5\overline{)17.0} \\ \underline{15} \\ 2\,0 \\ \underline{2\,0} \end{array}$$

Calculator Division

Most calculators display a remainder as a decimal fraction. When you work with dollars, a decimal fraction represents cents.

For some problems, if the calculator shows a decimal remainder, you need to round to the next whole number.

EXAMPLE 1   Lena is packing 16 books in each box. How many boxes will she need for 189 books?

To solve, divide 189 by 16. The whole number part of the answer tells us that 11 boxes will be filled with books ($11 \times 16 = 176$). There is a remainder, so one more box will be needed. (That box will have $189 - 176 = 13$ books.)

Press Keys	Display Reads
1 8 9	189.
÷	189.
1 6	16.
=	11.8125

11 boxes

One more box is needed to pack the remaining books.

ANSWER: 12 boxes (11 + 1)

Use your calculator to solve each problem.

1. A schoolbus can transport 80 students. How many buses are needed to transport 475 students?

2. A computer disk can hold electronic files for 35 photographs. How many computer disks are needed to hold electronic files for 450 photographs?

Finding the Value of a Remainder

In some problems, you want to know the value of a remainder.

EXAMPLE 1 Jason needs to haul 94 cubic yards of dirt by the end of the day. His truck can carry a maximum of 12 cubic yards. Assuming he fills his truck when possible, how much dirt will Jason carry on his final load? He can follow these three steps to find out.

STEP 1 Divide 94 by 12.

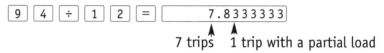

7 trips   1 trip with a partial load

The whole-number part of 7.8333333 means that Jason will make 7 trips, each with a full load of 12 cubic yards, and 1 final trip with a partial load.

STEP 2 Find the amount of dirt hauled in 7 trips with full loads.

$$\boxed{7}\;\boxed{\times}\;\boxed{1}\;\boxed{2}\;\boxed{=}\;\boxed{8\,4\,.}\;\text{cubic yards}$$

STEP 3 Subtract to find the amount of dirt in the partial load.

$$\boxed{9}\;\boxed{4}\;\boxed{-}\;\boxed{8}\;\boxed{4}\;\boxed{=}\;\boxed{1\,0\,.}\;\text{cubic yards}$$

EXAMPLE 2 Divide 1,293 by 172. Find the value of the remainder.

STEP 1 Divide 1,293 by 172.

$$\frac{1{,}293}{172} = 7.5174418$$

$$\boxed{7.5174418}$$

↑ whole number part of answer

STEP 2 Multiply the divisor times the whole-number part of the answer.

$7 \times 172 = 1{,}204$

STEP 3 Subtract your result from the original dividend. This answer is the remainder.

$1{,}293 - 1{,}204 = 89$

ANSWER: 7 r89

Complete the steps in the following problem.

1. Find the whole number remainder for the problem 598 ÷ 27.

> **STEP 1** Divide 598 by 27
> Displayed answer: _____
>
> **STEP 2** Multiply the whole-number part of the answer found in Step 1 by 27.
> Displayed answer: _____
>
> **STEP 3** Subtract the answer found in Step 2 from 598.
> Displayed answer: _____

ANSWER: _____ r _____

Use your calculator. Write each answer as a whole number and a remainder.

2. 258 ÷ 17 = _____ r _____

3. 494 ÷ 53 = _____ r _____

4. 749 ÷ 41 = _____ r _____

5. 980 ÷ 25 = _____ r _____

Solve each word problem below.

6. Jason agreed to remove 114 cubic yards of gravel from a building site. Assuming he again carries 12 cubic yards each full load, how many cubic yards of gravel will Jason carry on his final load?

7. On a cross-country trip, Shari drove 350 miles each day except her final day. If she drove a total of 3,185 miles, determine the number of miles Shari drove the final day of the trip.

Discovery: Once in a while, you may accidentally try to do a calculation that your calculator is unable to do. The calculator will then display an **error symbol**—an E on most calculators. Here are two examples.

- **Overflow error:** Multiplying numbers that give a product that is too large for the display. Example: 18,500 × 9,650

1.7852500 E

- **Division by 0 error:** Trying to divide by 0. You cannot divide by 0. Example: 45 ÷ 0

0. E

If an error symbol appears on your calculator display, press a **clear** key and redo your calculation. If you still get an error symbol, do the problem with paper and pencil.

Puzzle Power

 By now you're probably pretty good at using a calculator. Review your skills by working these problems. Write your answers in the puzzle.

Across

1. $37 \times 19 =$

3. the number of cents more than $6.00 in the amount $6.20

4. $451 + 396 + 84 =$

6. $36.27 rounded to the nearest ten dollars

7. the best estimate of the product 39×21: 700, 800, or 900

9. $\$30.25 \times 4 \times 3 =$

10. the whole number part of the quotient $143 \div 14$

Down

1. $4,284 \div 6 =$

2. $702 \div 18 =$

3. $20.87 rounded to the nearest dollar

5. $1,241 - 529 - 362 =$

7. 827 rounded to the nearest ten

8. the digits displayed when you enter 19 cents on your calculator

9. the best estimate of the quotient $892 - 31$: 30, 40, or 50

DECIMALS AND FRACTIONS

Working with decimals comes easily on a calculator. You have already used the decimal point key $\boxed{\cdot}$, tenths, and hundredths when you solved problems involving money. Fractions are a little trickier to represent on a calculator, but you'll learn an easy way to work with them. You will also learn an easy way to enter mixed numbers, such as $4\frac{1}{3}$ or $12\frac{5}{8}$, using the $\boxed{=}$ key.

Introducing Decimals

What's the most common use of decimals? You're correct if you answer, "Money!"

• Cents are the **decimal** part of a dollar.

If you enter keystrokes such as $\boxed{1}$ $\boxed{1}$ $\boxed{\div}$ $\boxed{8}$ $\boxed{=}$, your calculator displays $\boxed{1.375}$, which shows the second most common use of decimals.

• The remainder is often written as a decimal.

Reading Decimals

To get yourself ready to work with decimal place values, review the meaning of each simple decimal below.

Decimal	Value	Meaning
0.1	one-tenth	1 part out of 10 parts
0.01	one-hundredth	1 part out of 100 parts
0.001	one-thousandth	1 part out of 1,000 parts
0.0001	one ten-thousandth	1 part out of 10,000 parts
0.00001	one hundred-thousandth	1 part out of 100,000 parts
0.000001	one-millionth	1 part out of 1,000,000 parts

EXAMPLE   Read the name of a decimal.

		Decimal 1 0.042	**Decimal 2** 0.0070
STEP 1	Identify the place value of the digit farthest to the right.	thousandths	ten-thousandths
STEP 2	Read the number to the right of the decimal point. Ignore zeros at the left.	forty-two	seventy
STEP 3	Read the number and the place value together.	**forty-two thousandths**	**seventy ten-thousandths**

Mixed Decimals

A **mixed decimal** is a whole number plus a decimal. The number 7.18 is a mixed decimal. The amount $3.99 is also a mixed decimal.

mixed decimal = whole number + decimal

When reading a mixed decimal, read the decimal point as the word *and.* Do not say *and* in any other part of the number. For example, 107.18 is read "one hundred seven *and* 18 hundredths." The money amount $1,075.99 is read "one thousand seventy-five dollars *and* 99 cents."

Write words to express the value of each decimal below.

1. 0.05 _____

2. 0.5 _____

3. 0.000005 _____

4. 0.005 _____

5. 0.00005 _____

6. 0.0005 _____

Determine the value of each displayed number below. Choose each answer from the choices given.

7. | 0.206 |
   a. 26 hundredths
   b. 206 hundredths
   c. 206 thousandths

8. | 0.047 |
   a. 47 tenths
   b. 47 hundredths
   c. 47 thousandths

9. | 0.7 |
   a. 7 tenths
   b. 7 hundredths
   c. 7 thousandths

10. | 0.15 |
    a. 15 hundredths
    b. 15 thousandths
    c. 150 ten-thousandths

11. | 0.009 |
    a. 9 tenths
    b. 9 thousandths
    c. 9 ten-thousandths

12. | 0.08 |
    a. 8 tenths
    b. 8 hundredths
    c. 80 hundredths

Use your calculator to solve each of the following division problems. Choose the correct way to say the answer to the problem.

13. 1 ÷ 8 =
    a. 1 and 25 tenths
    b. 125 thousandths
    c. 1 and 25 hundredths

14. 9 ÷ 1 0 0 =
    a. 9 tenths
    b. 9 hundredths
    c. 9 thousandths

15. 3 0 ÷ 8 =
    a. 375 hundredths
    b. 3 and 75 hundredths
    c. 3 and 75 thousandths

16. 5 ÷ 1 6 =
    a. 3,125 hundredths
    b. 3,125 thousandths
    c. 3,125 ten-thousandths

17. 2 0 ÷ 8 0 =
    a. 25 tenths
    b. 2 and 5 tenths
    c. 25 hundredths

18. 2 4 5 ÷ 5 6 =
    a. 4 and 375 thousandths
    b. 5,375 ten-thousandths
    c. 4,375

Rounding Decimals

For most problems, you will only need one or two decimal places.

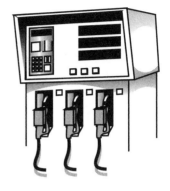

EXAMPLE Gasoline prices are always given to three decimal places. If the price per gallon is $1.489, how much do 16 gallons of gas cost?

$1.489 × 16 = $23.824

This answer rounds to $23.82.

In many cases, a rounded answer will do. You can use the sign ≈ to mean "is approximately equal to."

$1.489 × 16 = $23.824 ≈ **$23.82**

Rounding a Decimal

Steps for rounding decimals:

STEP 1 Find the digit you wish to round to, and underline that digit.

STEP 2 Look at the digit to the right of the underlined digit.

- If the digit to the right is greater than or equal to 5, add 1 to the underlined digit.

- If the digit to the right is less than 5, leave the underlined digit as it is.

- Discard all digits to the right of the underlined digit.

Rounded to the Tenths Place

tenths place

2.9̲82 ≈ 0.4

8 is greater than 5, so 2.982 rounds to 3.0

tenths place

4.6̲3 ≈ 4.6

3 is less than 5, so 4.63 rounds to 4.6

Rounded to the Hundredths Place

hundredths place

0.69̲4 ≈ 0.69

4 is less than 5, so 0.694 rounds to 0.69

hundredths place

7.03̲5 ≈ 7.04

5 is equal to 5, so 7.035 rounds to 7.04

Round each amount below to the nearest ten cents. Circle one of the two answer choices.

1. $0.27: $0.20   or   ($0.30)

2. $0.95: $0.90   or $1.00

3. $7.93: $7.90   or $8.00

4. $0.62: $0.60   or $0.70

5. $3.55: $3.50   or $3.60

6. $4.05: $4.00   or $4.10

Round each number below to the place value indicated.

	Tenths	Hundredths			Hundredths	Thousandths
7. 3.457	3.5	3.46	8.	8.0073	8.01	8.007
9. 14.382	_____	_____	10.	24.0049	_____	_____
11. 38.945	_____	_____	12.	90.8307	_____	_____
13. 40.050	_____	_____	14.	72.3846	_____	_____

 Use your calculator to divide. Write both the displayed answer and the rounded answer.

Round each answer to the tenths place.

15. $16 \div 13 = 1.2307692 \approx 1.2$ $11 \div 8 =$ $24 \div 7 =$

      ⤴ less than 5 $\approx$ _____ $\approx$ _____

Round each answer to the hundredths place.

16. $31 \div 16 = 1.9375 \approx \$1.94$ $15 \div 8 =$ $\$17 \div 9 =$

      ⤴ greater than 5 $\approx$ _____ $\approx$ _____

Round each answer to the thousandths place.

17. $23 \div 16 = 1.4375 \approx 1.438$ $45 \div 22 =$ $16 \div 7 =$

      ⤴ equal to 5 $\approx$ _____ $\approx$ _____

Solve each word problem below.

18.   A jeweler wishes to cut a 33-inch long gold wire into 7 equal pieces. To the nearest hundredth of an inch, how long should each of the 7 pieces be?

19.   One inch equals 2.54 centimeters. How many centimeters are equal to the length of 1 yard (36 inches)? Express your answer to the nearest tenth of a centimeter.

Terminating and Repeating Decimals

Decimals result each time you

- divide one number by another and get a remainder

  [6] [÷] [4] [=] [1.5]

- divide a smaller number by a larger number

  [4] [÷] [6] [=] [0.6666666]

For each type of division, the result is said to be either a **terminating decimal** or a **repeating decimal.**

Terminating Decimals

A terminating decimal has a limited number of digits.

EXAMPLES

One decimal digit
$3 \div 5 = 0.6$

[3] [÷] [5] [=] [0.6]

Two decimal digits
$7 \div 4 = 1.75$

[7] [÷] [4] [=] [1.75]

Three decimal digits
$21 \div 8 = 2.625$

[2] [1] [÷] [8] [=] [2.625]

Repeating Decimals

A repeating decimal has a never-ending, repeating pattern of one or more digits. Look at these examples.

EXAMPLES

Single repeating digit $4 \div 9 = 0.444 \ldots$

[4] [÷] [9] [=] [0.4444444]

Two repeating digits $144 \div 66 = 2.1818 \ldots$

[1] [4] [4] [÷] [6] [6] [=] [2.1818181]

Three repeating digits $148 \div 999 = 0.148148 \ldots$

[1] [4] [8] [÷] [9] [9] [9] [=] [0.1481481]

Rounding Versus Truncating

Notice that a repeating digit pattern continues forever! For this reason, a calculator gives an approximate answer to any division problem that results in a repeating decimal.

Try this division problem on your calculator: $2 \div 3 =$

- If your display shows 0.6666666, your calculator **truncates** (drops) any digits that won't fit on your display.

- If your display shows 0.6666667, your calculator **rounds** the answer to the final decimal place on your calculator display.

Most four-function calculators show truncated decimals. Other calculators, including those designed for use in business or science, show rounded values for decimals. See if anyone in your class has a rounding calculator.

. . . .

Use pencil and paper to perform each division below. Indicate with a check (✓) the type of decimal answer you obtain.

1. $4\overline{)5}$ _____ terminating
   _____ repeating

2. $3\overline{)1}$ _____ terminating
   _____ repeating

3. $33\overline{)5}$ _____ terminating
   _____ repeating

4. $8\overline{)7}$ _____ terminating
   _____ repeating

Use your calculator to perform each division below. Below each problem, indicate the pattern of repeating digits. Problem 5 is completed as an example.

5. $17 \div 11 = 1.5454545$

   repeating digits: 54

6. $4 \div 3 =$

   repeating digits:

7. $23 \div 9 =$

   repeating digits:

8. $8 \div 33 =$

   repeating digits:

9. Some answers do not terminate or show a repeating pattern in the first eight digits. Try $3 \div 7$ and $5 \div 17$.

Recognizing Patterns

As you saw on pages 49 and 50, repeating digits form a **pattern.** When you recognize a pattern, you can predict what will come next.

In your work with math, it will help to look for patterns. This page gives you practice with patterns in shapes, words, letters, and numbers.

See if you can draw the next shape in each pattern below. The first one is done for you.

1.   2.

3.   4.

Try to finish these familiar word patterns.

5. ten, nine, _____

6. first, second, _____

7. small, smaller, _____

8. good, better, _____

What is the next number or letter in each series below?

9. 1, 3, 6, 10, 15, _____

10. A, Z, B, Y, C, _____

11. 6, 15, 7, 14, 8, _____

12. A, C, F, J, O, _____

 For each problem, use your calculator to do the first three problems in each column. Then, using the answer pattern as a guide, predict the answer to each starred (*) problem.

13. $1001 \times 46 =$
    $1001 \times 52 =$
    $1001 \times 28 =$
    $*1001 \times 34 =$
    $*1001 \times 16 =$

14. $10 \div 9 =$
    $11 \div 9 =$
    $12 \div 9 =$
    $*13 \div 9 =$
    $*15 \div 9 =$

15. $37 \times 6 =$
    $37 \times 9 =$
    $37 \times 12 =$
    $*37 \times 15 =$
    $*37 \times 21 =$

16. $12 \div 33 =$
    $13 \div 33 =$
    $14 \div 33 =$
    $*15 \div 33 =$
    $*17 \div 33 =$

Recognizing Key Words

....

When you read a word problem, look for **key words** that tell you whether the problems calls for addition, subtraction, multiplication, or division. Some of the most familiar key words are listed below.

Some Word Problem Key Words

Addition	*sum, and, more, in all*
Subtraction	*less than, difference, lost*
Multiplication	*times, total, per*
Division	*equal, each, average*

 In each problem below, circle the symbol standing for the key word that you recognize. Then use your calculator to solve the problem.

1. Angie wants to save $510 to buy a new stereo. How much must she save each month if she plans to buy the stereo on one year's time? (one year = 12 months)

    $+$ $-$ $\times$ $\div$ Answer: _____

2. When she bought her house, Karla paid $148,795. She recently sold this house for $17,980 more than she paid for it. Given these figures, determine the sale price of the house.

    $+$ $-$ $\times$ $\div$ Answer: _____

3. When he started his diet, Lucas weighed 236 pounds. During the first 3 months, he lost 17 pounds. Now, 6 months later, Lucas weighs 199 pounds. How much has Lucas lost in all?

    $+$ $-$ $\times$ $\div$ Answer: _____

4. Hanna's car gets 22 miles per gallon in city driving and 28 miles per gallon on the highway. How many miles of city driving can Hanna get per 16 gallons of gas?

    $+$ $-$ $\times$ $\div$ Answer: _____

5. George and four friends are going camping. They've agreed that each of them should carry the same amount of weight while hiking. If the gear weighs 142 pounds, what weight should each of them carry?

    $+$ $-$ $\times$ $\div$ Answer: _____

Adding Two or More Decimals

When you use paper and pencil to add, the first step is to line up the decimal points. Your calculator does that step automatically. To catch keying errors, keep your eye on the number of decimal places in the answer.

EXAMPLE 1 $12.047 + 9.36 =$

To add 12.047 and 9.36 on your calculator, press the keys as shown.

Press Keys	**Display Reads**
1 2 · 0 4 7	12.047
+	12.047
9 · 3 6	9.36
=	21.407

ANSWER: 21,407

Discovery: When you use paper and pencil to add or subtract decimal numbers, you may add a placeholding zero.

When you use a calculator, you do not have to enter placeholding zeros.

Pencil and Paper

$$12.047$$
$$\underline{+\ 9.360} \quad \leftarrow \quad \text{placeholding 0}$$
$$21.407$$

EXAMPLE 2 $\$5.60 + \$9 + \$7.82 =$

To add $5.60, $9, and $7.82, press the keys as shown.

Press Keys	**Display Reads**
5 · 6 0	5.6
+	5.6
9	9.
+	14.6
7 · 8 2	7.82
=	22.42

ANSWER: $22.42

Discovery: When you add a whole number to a decimal, you do not need to enter a decimal point to the right of the whole number. The calculator does this for you.

Practice keeping your eye on the number of decimal places so you can catch keying errors. Complete each problem below by placing a decimal point in the answer. If you find it helpful, rewrite each problem with one number above the other, lining up the decimal points.

1. $0.56 + 0.8 = 1\ 3\ 6$ $0.7 + 0.29 = 9\ 9$ $2.4 + 0.75 + 0.8 = 3\ 9\ 5$

2. $0.07 + 0.09 = 1\ 6$ $1.6 + 0.9 = 2\ 5$ $3.06 + 8 + 0.59 = 1\ 1\ 6\ 5$

Calculate an exact answer to each problem below. Then compute an estimate, following these guidelines.
- Round numbers smaller than 10 to the nearest whole number.
- Round numbers between 10 and 99 to the nearest ten.
- Round numbers of 100 or more to the nearest hundred.

	Exact	Estimate	Exact	Estimate	Exact	Estimate
3.	8.3 + 5.6 ―――― 13.9	8 + 6 ―――― 14	9.7 + 6.5 ――――		6.3 + 7.1 ――――	
4.	25.8 + 13.6 ――――		75.60 + 24.83 ――――		27.85 + 12.4 ――――	
5.	328.65 211.06 + 123 ――――		547.69 243.63 + 181.8 ――――		900 375.8 + 245.18 ――――	

Calculate the exact answer to each problem. Round each answer to the nearest whole number (or dollar).

6. $21.8 + 14.7 =$ $57.14 + 37.5 =$ $\$154.85 + \$49.50 =$

7. $250 + 136.8 =$ $231.8 + 85.95 =$ $\$34.54 + \$23.50 =$

Use your calculator to solve the problem below.

8. Find the combined length in inches of the three spacers pictured at the right.

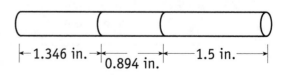

Subtracting Two or More Decimals

You subtract decimals in the same way you subtract whole numbers. Be sure to enter the numbers in the correct order! In decimal subtraction, as in decimal addition, the calculator lines up the decimal points for you.

EXAMPLE 1 $9.1 - 4.76 =$

To subtract 4.76 from 9.1 on your calculator, press the keys as shown.

Press Keys	Display Reads
9 . 1	9.1
−	9.1
4 . 7 6	4.76
=	4.34

ANSWER: 4.34

EXAMPLE 2 $15 - 3.5 - 8.375 =$

To subtract 3.5 and 8.375 from 15, press the keys shown.

Press Keys	Display Reads
1 5	15.
−	15.
3 . 5	3.5
−	11.5
8 . 3 7 5	8.375
=	3.125

ANSWER: 3.125

EXAMPLE 3   Fran must cut a dowel so that its finished length is 8.765 inches. If she starts with a 10-inch-long dowel, how much will Fran need to remove?

To solve, subtract 8.765 from 10.

Press Keys	Display Reads
1 0	10.
−	10.
8 . 7 6 5	8.765
=	1.235

ANSWER: 1.235 inches

When you subtract decimals, read the problem carefully to decide which number to enter first. (You will enter the larger decimal in the calculator first.)

EXAMPLE 4 Sam Rogers is 1.46 meters tall, and his brother Ken is 1.7 meters tall. Which brother is taller and by how much?

STEP 1 Compare the number of decimal places. If there are not the same number of places, add placeholder zeros to give them the same number of places.	1.7	1.46
	1.70 larger	1.46
STEP 2 Subtract the smaller number from the larger.	1.70 – 1.46 = 0.24	

ANSWER: Ken Rogers is 0.24 meters taller.

· ·

Circle the larger decimal in each pair below.

1. 0.81 or 0.79   2. 0.102 or 0.11   3. 0.035 or 0.0093

Find the difference between each pair of decimals below. (If you accidentally subtract the larger from the smaller, your calculator will display a minus sign next to the answer.)

4. 0.36 and 0.59   0.93 and 0.755   0.405 and 0.386

5. 0.09 and 0.134   0.8 and 0.79   0.354 and 0.38

Calculate the answer to each problem.

6.   0.298
   – 0.145

7. $13.45
   –   8.80

8. $25 – $12.45 – $11.85 =

9. When Shannon had the flu, her temperature rose as shown below. How much higher is her fever temperature than normal?

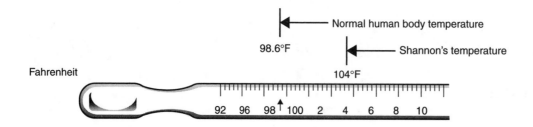

Decimal Addition and Subtraction Problems

 For problems 1–3, substitute whole numbers for decimals and estimate an answer. Use your estimate as a guide to choose the exact answer from the choices given. (Estimate, don't calculate!)

1. At a picnic, two tables were placed end to end. If the first table is 2.89 meters long and the second is 1.94 meters long, what is the combined length in meters of the two tables?

    (Substitute 3 for 2.89 and 2 for 1.94.)

    a. 3.13　　　　　　　**b.** 4.83　　　　　　　**c.** 6.73

2. Ming is trying to decide which of two pork roasts to buy. One weighs 4.79 pounds, and the other weighs 6.2 pounds. How much heavier is the larger roast than the smaller one?

    (Substitute 5 for 4.79 and 6 for 6.2.)

    a. 1.41　　　　　　　**b.** 2.61　　　　　　　**c.** 3.01

3. Jason Sports is advertising that ski jacket prices have been drastically reduced from $61.89 to $39.99. How much of a savings is this advertised price reduction?

    (Substitute $60 for $61.89 and $40 for $39.99.)

    a. $14.80　　　　　　**b.** $16.90　　　　　　**c.** $21.90

 For problems 4–6, use your calculator to compute an exact answer. Then estimate an answer as a check against keying errors.

4. With his football uniform on, Rocky can run the 100-yard dash in 14.26 seconds. Wearing only running clothes, he can run the same distance in 11.89 seconds. How much faster can Rocky run the 100-yard dash when he's not wearing football gear?

    _____
    exact

    _____
    estimate

5. Rose switched to super unleaded gas and found that her gas mileage went from 27.8 miles per gallon to 30.2 miles per gallon. By how many miles per gallon did the super unleaded gas improve her mileage?

_____
exact

_____
estimate

6. Following the weekend sale, The Gift Shoppe raised the prices of all vases by $2.29. What would be the new price of a vase that had been priced at $11.88 during the sale?

_____
exact

_____
estimate

For problems 7–9, look at the list or drawing to the right of each problem to find necessary information.

7. As shown on the map at the right, Leslie lives almost halfway between the theater and the swimming pool. By how many miles is Leslie closer to one than the other?

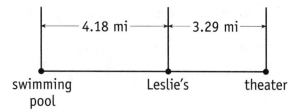

|←— 4.18 mi —→|←— 3.29 mi —→|

swimming   Leslie's   theater
pool

_____   _____
exact   estimate

8. Gloria keeps a record of her gasoline purchases. Part of that record is shown at the right. How many more gallons of gas did Gloria purchase on February 28 than on January 24?

_____   _____
exact   estimate

Date	Gallons
1/9	21.4
1/24	18.9
2/11	17.9
2/28	20.7

9. Expressing your answer as a decimal fraction, how much longer is the total length of the two shorter bolts than the length of the longest bolt.

_____   _____
exact   estimate

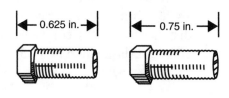

|←— 0.625 in. →| |←— 0.75 in. →|

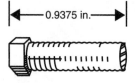

|←— 0.9375 in. —→|

Multiplying Two or More Decimals

You can use your calculator to multiply a decimal times a whole number or to multiply two or more decimals.

As in whole number multiplication, enter each number into the calculator and press ×. Press = only once, after the final number is multiplied.

EXAMPLE 1 $6.84 \times 3.7 =$

	Press Keys	**Display Reads**
To multiply 6.84 times 3.7 on your calculator, press the keys as shown.	6 · 8 4	6.84
	×	6.84
	3 · 7	3.7
	=	25.308

ANSWER: 25.308

EXAMPLE 2 Multiply $5.06 \times 4 \times 0.088$. Round the answer to the nearest cent.

	Press Keys	**Display Reads**
To solve, press keys as shown.	5 · 0 6	5.06
	×	5.06
	4	4.
	×	20.24
	· 0 8 8	0.088
	=	1.78112

ANSWER: $1.78

Discovery: When you compute with paper and pencil, you total the number of decimal places in the problem to place the decimal point in the answer.

The calculator correctly places the decimal point for you.

Pencil and Paper

6.84	two decimal places
× 3.7	+ one decimal place
25.308	three decimal places
3 2 1	

Complete the following problems by correctly placing a decimal point
in each answer. You will be able to catch keying errors on your
calculator if you have a sense of where decimal points should be placed.

1. 7.8	25.7	8.04	13.1	12.4
× 6	× 0.3	× 0.7	× 5.5	× 0.062
4 6 8	7 7 1	5 6 2 8	7 2 0 5	7 6 8 8

Calculate the exact answer. Then estimate an answer to check against
possible keying errors.

Exact	Estimate	Exact	Estimate	Exact	Estimate
2. 6.8	7	8.12		4.03	
× 4.2	× 4	× .94		× 2.1	
	28				

3. 10.3	12.9	15.4
× 5.9	× 1.09	× 2.19

Use your calculator to solve each of the following problems. Round
each answer to the tenths place.

4. $7.5 \times 3.2 \times 6 =$ $9.6 \times 8.3 \times 5.7 =$

5. $21.4 \times 3.1 \times 5.2 =$ $31.5 \times 5.5 \times 3.2 =$

Round each answer to the nearest cent.

6. $\$2.75 \times \$8.80 \times \$4.60 =$ $\$15.45 \times \$3.50 \times \$5.25 =$

7. To the nearest cent, what will Beth pay for 16.4 gallons
of gas bought at the Gas City pump price shown
at the right?

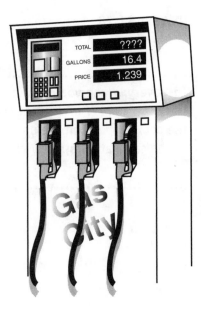

Dividing Decimals

As the following examples show, you can use your calculator to divide a decimal by a whole number or to divide one decimal number by another.

As in whole number division, your fist step is to correctly identify the dividend (the number *being divided*) and the divisor (the number you are *dividing by*).

EXAMPLE 1 Divide 7.894 ÷ 4.

To solve, press the keys as shown.

Press Keys	Display Reads
[7] [·] [8] [9] [4]	7.894
[÷]	7.894
[4]	4.
[=]	1.9735

ANSWER: 1.9735

> **Discovery:** A calculator carries out division until there is no remainder—or until the display is full. For this reason, the answer may contain more decimal places than the dividend (the number being divided). See Examples 1 and 2.

EXAMPLE 2 Divide $154.85 by 6.8 and round the answer to the nearest cent.

To solve, press the keys as shown.

Press Keys	Display Reads
[1] [5] [4] [·] [8] [5]	154.85
[÷]	154.85
[6] [·] [8]	6.8
[=]	22.772058

ANSWER: $22.77

EXAMPLE 3 One foot (12 inches) is the same as 30.48 centimeters. How many centimeters are in 1 inch?

To solve, divide 30.48 by 12, as shown.

Press Keys	Display Reads
[3] [0] [·] [4] [8]	30.48
[÷]	30.48
[1] [2]	12.
[=]	2.54

ANSWER: There are 2.54 centimeters in 1 inch.

In the following problems, identify the dividend (the number *being* divided) and the divisor (the number you are *dividing by*). Circle the number you enter into your calculator first. Do not solve.

1. $6.54 \div 3.1 =$ dividend _____

divisor _____

2. 8 divided by 2.7 dividend _____

divisor _____

Circle the answer choice that is the best estimate for each problem below. Estimating answers can help you to catch keying errors on the calculator.

3. $47.9 \div 6.1 =$ **a.** 6   **4.** $4.1\overline{)33.7}$ **a.** 8
                      b. 8   **b.** 11
                      c. 10   **c.** 16

5. $104.8 \div 11.6 =$ **a.** 5   **6.** $8.01\overline{)57.75}$ **a.** 3
                      b. 10   **b.** 5
                      c. 15   **c.** 7

Round each answer to the tenths place.

7. $363.2 \div 8.1 =$ $76.26 \div 3.14 =$ $12\overline{)604.86}$ $13.8\overline{)130.41}$

Round each answer to the hundredths place.

8. $78 \div 2.3 =$ $\$24.71 \div 8 =$ $6\overline{)\$142.88}$ $3.7\overline{)8.93}$

9. Sherry bought a roast and weighed it on a scale. Fill in the $ PER LB amount to show how much Sherry is paying per pound. Enter your answer to the nearest cent.

Scale	
Total Price	$33.17
5.87 LB	$ __.__ __
Weight	$ PER LB

Fill in
← this
amount.

Estimating with Multiplication and Division

Circle the choice that best describes each correct answer.
Hint: Estimate by substituting whole numbers for decimals.

1. Working as a salesman, Geoff earned $184.28 in commissions
   Saturday during one 9-hour shift. On the average, how much did
   Geoff earn each hour in commissions?

   a. between $16 and $18   **b.** between $18 and $20   **c.** between $20 and $22

2. In the metric system, weight is measured in units called grams. If
   28.4 grams is equal to 1 ounce, how many grams does an 11-ounce
   steak weigh?

   a. between 150 and 200   **b.** between 250 and 350   **c.** between 350 and 450

3. Joanne earns $11.50 an hour as a part-time hairstylist. At this rate,
   how much will Joanne earn in 7.25 hours?

   a. between $58 and $77   **b.** between $76 and $96   **c.** between $97 and $116

4. The distance between Mustafa's house and his office is 2.85 miles.
   He walks this distance twice each day, Monday through Friday.
   How many total miles does Mustafa walk each week between his
   house and work?

   a. between 12 and 18   **b.** between 19 and 25   **c.** between 26 and 32

5. Kerri paid $16.53 for a package of coffee that was marked down
   from $24.95. The weight of the package was 5.14 pounds. How
   much did Kerri pay per pound for this coffee?

   a. between $3 and $4   **b.** between $5 and $6   **c.** between $7 and $8

**First use your calculator to compute an exact answer. Then estimate
answers as a check against keying errors.**

6. At Big Bear Foods, Vicki paid $3.13 for a sack of bananas. If the
   sack weighed 2.9 pounds, how much did Vicki pay per pound?

   _____
   exact

   _____
   estimate

7. Yoshi's Import Foods received a box weighing 59.8 pounds that was filled with jars of pickled vegetables. If the box contains 22 jars, what is the weight of each jar to the nearest tenth of a pound? (Ignore the weight of the box itself.)

_____
exact

_____
estimate

8. As a decimal fraction, $\frac{15}{16}$ of an inch is written as 0.9375 inch. How wide would a group of eight paperback books be if each book is $\frac{15}{16}$ of an inch wide? Write your answer as a mixed decimal.

_____
exact

_____
estimate

9. Water weighs approximately 8.3 pounds per gallon. Determine the weight of water in a hot water heater that contains 48 gallons when filled.

_____
exact

_____
estimate

 Problems 10–13 refer to the following story. Use your calculator to compute answers to each question.

Anna is trying to decide which of two jobs to take. If she works for Data Systems, she'll be paid $9.44 an hour and work a standard 40-hour shift. However, a friend who works there has told Anna that Data Systems almost always offers 5 hours of overtime work each week to each employee. Data Systems pays an overtime pay rate of 1.5 times normal hourly wage.

Anna has also been offered a job at Brinson Electronics. At Brinson, Anna would be paid a weekly salary of $425.50 for 40 hours work. Brinson Electronics does not normally offer overtime work hours.

10. Not counting overtime, how much would Anna's weekly salary be at Data Systems?

11. Determine how much Anna would make for each hour of overtime at Data Systems.

12. Working 5 hours of overtime each week, how much total pay could Anna earn weekly at Data Systems?

13. Figure out how much Anna would make per hour if she decides to choose the job at Brinson Electronics.

Application: Comparing Costs of Child Care

 Child care is an expense for many families. One important step in choosing a child-care center is to compare the costs of different centers. The exercise on this page deals with this type of comparison.

Sheri and Roberto Robertson need child care for their daughter Ashley for three months in the summer—June, July, and August. They want Ashley to eat lunch each day at the center.

As shown below, Sheri made a list of the costs of sending Ashley to each of the two centers in their community. Ashley would be using transportation at either center.

	Super Kids Care	Huggy Bear Center
Registration Fee (one-time fee, paid the first month)	$35.00	$65.00
Child care	$255.50/month	$15.85/day*
Transportation (only charged for days of attendance)	$6.75/day*	$4.50/day*

*Figure that every month has 20 working days.

 Use your calculator to help determine each of the costs indicated below.

	Super Kids Care	Huggy Bear Center
1. Total costs for June	_____	_____
2. Total costs for July and August together	_____	_____
3. Total costs for three-month summer attendance	_____	_____

4. What is the difference in the three-month costs between the two choices that Sheri has listed?

Reviewing Fractions

A **fraction** stands for part of a whole. A fraction is written as one number over another.

$\frac{3}{4}$ ← numerator
← denominator

The **numerator** tells how many parts you have.
The **denominator** tells how many parts one whole is divided into.

Types of Fractions

Fractions are written in one of three ways.

- As a **proper fraction** in which the top number is always less than the bottom number.

  EXAMPLE $\frac{1}{3}, \frac{2}{3}$

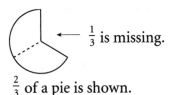

← $\frac{1}{3}$ is missing.

$\frac{2}{3}$ of a pie is shown.

- As an **improper fraction** in which the top number is the same as or greater than the bottom number.

  EXAMPLE $\frac{5}{3}$

$\frac{5}{3}$ is $\frac{2}{3}$ larger than 1 whole $\left(\frac{3}{3}\right)$.

$\frac{5}{3}$ pies are shown.

- As part of a **mixed number,** which is the sum of a whole number and a proper fraction.

  EXAMPLE $1\frac{3}{4}$

$1 + \frac{3}{4} = 1\frac{3}{4}$

$1\frac{3}{4}$ pies are shown.

Changing Fractions to Decimals

You can use your calculator to change a fraction to a decimal.

- To change a proper or improper fraction to a decimal, divide the numerator by the denominator.

EXAMPLE 1 Change $\frac{9}{16}$ to a decimal.

Divide 9 by 16 by pressing the keys shown.

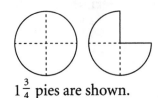

Press Keys	Display Reads
9	9.
÷	9.
1 6	16.
=	0.5625

ANSWER: $\frac{9}{16} = 0.5625$

- To tell which is greater, a fraction or a decimal, change the fraction to a decimal.

EXAMPLE 2 Which is greater, $\frac{11}{16}$ or 0.65?

		Press Keys	**Display Reads**
STEP 1	Write $\frac{11}{16}$ as a decimal. As shown at the right, $\frac{11}{16} = 0.6875$.	1 1	11.
STEP 2	To compare 0.6875 with 0.65, give each decimal the same number of decimal places. Do this by adding two placeholder zeros to 0.65.	÷ 1 6 =	11. 16. 0.6875

$$0.65 = 0.6500$$
$$\frac{11}{16} = 0.6875$$

ANSWER: $\frac{11}{16}$ **is greater** than 0.65 because 0.6875 is greater than 0.6500.

··

Use your calculator to change each fraction to a decimal. Then circle the greater number in each pair.

1. $\frac{3}{4}$ or 0.76 $\frac{5}{8}$ or 0.618   0.4 or $\frac{4}{15}$ 0.865 or $\frac{5}{7}$

Change both fractions to decimals and then compare. Circle the greater fraction.

2. $\frac{2}{3}$ or $\frac{3}{5}$ $\frac{9}{17}$ or $\frac{6}{13}$ $\frac{21}{32}$ or $\frac{2}{3}$ $\frac{8}{10}$ or $\frac{51}{64}$

3. Which of the following is the greater amount: $0.35 or $\frac{3}{8}$ of a dollar?

Solve each word problem below.

4. Cal needs to drill a hole that will allow a 0.2 inch-diameter wire to pass through. He wants the wire to fit the hole as tightly as possible. Which of the three drill bits will give the wire the tightest fit?

Bit	Diameter
A	$\frac{3}{16}$ inch
B	$\frac{7}{32}$ inch
C	$\frac{13}{64}$ inch

5. Will a bridge that has a load limit of $5\frac{2}{3}$ tons safely support a loaded truck that weighs 5.754 tons?

Calculations with Mixed Numbers

Fractions and mixed numbers can be added, subtracted, multiplied, and divided. Below are examples of paper-and-pencil solutions.

Addition	Subtraction	Multiplication	Division
$\frac{2}{3} = \frac{10}{15}$	$1\frac{1}{2} = \frac{3}{2} = \frac{9}{6}$	$2\frac{3}{4} \times \frac{7}{8}$	$\frac{2}{5} \div 2\frac{1}{4}$
$+\frac{3}{5} = +\frac{9}{15}$	$-\frac{5}{6} = -\frac{5}{6}$	$= \frac{11}{4} \times \frac{7}{8}$	$= \frac{2}{5} \div \frac{9}{4}$
$\frac{19}{15} = 1\frac{4}{15}$	$\frac{4}{6} = \frac{2}{3}$	$= \frac{77}{32}$	$= \frac{2}{5} \times \frac{4}{9}$
		$= 2\frac{13}{32}$	$= \frac{8}{45}$

Calculators and Fractions

Four-function calculators are not designed to do calculations like those above. Even though your calculator may not solve fraction problems simply and directly, you can still use it to solve fraction problems when a decimal answer is required or when you are given answer choices.

EXAMPLE 1 $2 \times 43\frac{5}{12} =$ Round the answer to the hundredths place.

		Press Keys	Display Reads
STEP 1	Enter the fraction.	5	5.
		÷	5.
		1 2	12.
STEP 2	Add the whole number.	+	0.4166666
		4 3	43.
STEP 3	Press the = key.	=	43.416666
STEP 4	Multiply by 2.	×	43.416666
		2	2.
STEP 5	Press the = key.	=	86.833332

ANSWER: 86.83

EXAMPLE 2 Matt bought $2\frac{1}{2}$ pounds of apples and $6\frac{2}{3}$ pounds of cherries. How much fruit did he buy?

    a. $9\frac{1}{6}$ lb   **b.** $9\frac{1}{2}$ lb   **c.** $9\frac{1}{3}$ lb

    STEP 1 Change the mixed numbers to decimals.

$$2\frac{1}{2} = \frac{1}{2} + 2 = \boxed{1}\ \boxed{\div}\ \boxed{2}\ \boxed{+}\ \boxed{2}\ \boxed{=}\ 2.5$$

$$6\frac{2}{3} = \frac{2}{3} + 6 = \boxed{2}\ \boxed{\div}\ \boxed{3}\ \boxed{+}\ \boxed{6}\ \boxed{=}\ 6.67$$

    STEP 2 Add the decimals.

$$2.5 + 6.67 = 9.17$$

    STEP 3 Convert the other mixed fractions and compare.

    a. $9\frac{1}{6} = \frac{1}{6} + 9 = 9.1666 \approx 9.17$

    b. $9\frac{1}{2} = \frac{1}{2} + 9 = 9.5\quad = 9.50$

    c. $9\frac{1}{3} = \frac{1}{3} + 9 = 9.333\ \approx 9.33$

ANSWER: 9.17 lb is the decimal approximation for $9\frac{1}{6}$ lb. The answer is choice **a. $9\frac{1}{6}$ lb**

• •

Use your calculator to solve each problem below. For problems 1–4, choose each answer from the choices given.

1. $\quad 4\frac{3}{8}$
    $+3\frac{5}{6}$

    a. $7\frac{43}{48}$
    b. $7\frac{11}{12}$
    c. $8\frac{5}{24}$

2. $\quad 4 \times 3\frac{2}{3} =$

    a. $14\frac{1}{3}$
    b. $14\frac{2}{3}$
    c. $14\frac{5}{6}$

3. $\quad 9\frac{6}{7}$
    $-3\frac{4}{5}$

    a. $6\frac{2}{35}$
    b. $6\frac{6}{35}$
    c. $6\frac{9}{35}$

4. $\quad 6\frac{4}{5} \div 3 =$

    a. $2\frac{1}{6}$
    b. $2\frac{4}{15}$
    c. $2\frac{5}{12}$

5. Linda paid \$17.72 for $12\frac{3}{4}$ pounds of peaches. To the nearest cent, how much is Linda paying per pound?

6. Noreen is buying $3\frac{3}{4}$ gallons of punch for a party. At \$2.95 a gallon, how much will she pay?

Calculations with Improper Fractions

Can you use your calculator to rewrite an improper fraction such as $\frac{281}{15}$ as a mixed number? The next example shows you how.

EXAMPLE Rewrite the improper fraction $\frac{281}{15}$ as a mixed number.

		Press Keys	**Display Reads**
STEP 1	Divide the numerator 281 by the denominator 15.	[2] [8] [1]	281.
		[÷]	281.
		[1] [5]	15.
		[=]	18.733333
STEP 2	Multiply the whole-number part by the denominator 15.	[1] [8]	18.
		[×]	18.
		[1] [5]	15.
		[+]	270.
STEP 3	Enter the original numerator 281, and subtract 270.	[2] [8] [1]	281.
		[−]	281.
		[2] [7] [0]	270.
		[=]	11.

ANSWER: For the mixed number, the whole number is part of your result from Step 1. The numerator is the result from Step 3, and the denominator is the denominator of the original improper fraction.

$$\frac{281}{15} = 18\frac{11}{15}$$

To check: $18\frac{11}{15} = 18 \cdot \frac{15}{15} + \frac{11}{15} = \frac{270}{15} + \frac{11}{15} = \frac{281}{15}$

 Use your calculator to rewrite each improper fraction as a mixed number.

1. $\frac{253}{7}$ $\frac{135}{19}$ $\frac{311}{27}$

2. $\frac{414}{17}$ $\frac{676}{23}$ $\frac{500}{11}$

3. $\frac{401}{16}$ $\frac{423}{95}$ $\frac{705}{112}$

4. $\frac{995}{23}$ $\frac{850}{73}$ $\frac{666}{11}$

Calculator Tic-Tac-Toe

Choose another student or friend to play calculator tic-tac-toe with you. Use pencils to mark the board lightly so that marks can be erased. You can play this game numerous times.

109.4	18	47.43
5.1	70.1	57.7
89.9	143.5	10

a. $35.8 + 97.3 =$

b. $41\frac{2}{5} + 28.7 =$

c. $88.9 - 31.2 =$

d. $64.9 - 49.8 =$

e. $97.8 + 31.4 + 28.2 =$

f. $57.8 + 46.3 + 39.4 =$

g. $100 - 57.3 - 26.8 =$

h. $200 - 62.7 - 27.9 =$

i. $11.9 \times 21.6 =$

j. $5.8 \times 5 \times 3.1 =$

k. $9\frac{3}{10} \times 5.1 =$

l. $46.92 \div 9.2 =$

m. $73.71 \div 9.1 =$

n. $7\frac{1}{5} \times 2.5 =$

o. $4\frac{3}{7} \times 7 =$

p. $3.5 \div 0.35 =$

Rules:

1. The goal of each player is to place his or her initials on a row of three squares. The row can be across, up and down, or on a diagonal.

2. Player A chooses a letter from *a* to *p* and then does the calculation. If the calculated answer is one of the numbers in a square, Player A places his or her initials on that square. It then becomes Player B's turn.

   If Player A chooses a problem whose answer is not in a square, Player A loses that turn and it becomes Player B's turn.

3. Players alternate turns until one player completes a row of three squares. That player is then the winner.

Hint: A player is allowed to make mental estimates of answers before doing a calculation. Using estimates, a player can first pick a square and then try to find a problem whose answer will match that square.

PERCENT

People use percent as part of daily life. We use percent when we calculate sales taxes, interest amounts, costs of sale items, and many other things. Two tools can make it easier to calculate with percent—your calculator and a memory device called the **percent circle.**

In this part of *Calculator Power,* you will use these two tools. On your calculator, you will use the percent key %, and on the next pages you will explore the **percent circle.**

Overview for Percent

The three main types of percent problems are listed below. They are based on the statement

25%   of $300   is $75.

Some **percent** of a **whole amount** is a **part** of that amount.

- Finding **part** of a whole

  What is 25% of $300?
  $75

- Finding what **percent** a part is of a whole

  What percent of $300 is $75?
  25%

- Finding a **whole** when a part of it is given
  If 25% of the price is $75, what is the total price? **$300**

In a percent problem, you are given two of these factors (*part, percent* or *whole*), and you must find the third.

The Percent Circle

On the pages ahead, we'll show you how a calculator can simplify solving each type of percent problem. On those pages we'll refer to a memory device shown below called the **percent circle.**

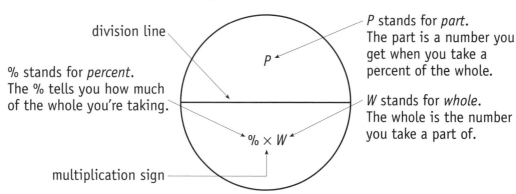

division line

% stands for *percent*.
The % tells you how much of the whole you're taking.

P stands for *part*.
The part is a number you get when you take a percent of the whole.

W stands for *whole*.
The whole is the number you take a part of.

multiplication sign

The following three examples show how the percent circle can help you remember how to find the part, the percent, or the whole.

EXAMPLE 1 **Finding the part**

If 17% of your $300 check is withheld for taxes, how much is withheld?

Cover the *P* (part), the number you are trying to find. The uncovered symbols tell you this is a *multiplication* problem.

$P = \% \times W$

$P = 17\% \times \$300 = \textbf{\$51}$

EXAMPLE 2 Finding the percent

In just over three months, Amy lost 40 pounds. If she originally weighed 200 pounds, what percent of her weight did she lose?

Cover the % (percent), the number you are trying to find. The uncovered symbols tell you this is a *division* problem.

$\% = P \div W$

$\% = 40 \div 200 =$ **20%**

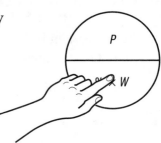

EXAMPLE 3 Finding the whole

When he bought a used piano, Willis paid $270 as a down payment. If this payment is 15% of the price of the piano, what was the full price?

Cover the W (whole), the number you are trying to find. The uncovered symbols tell you this is a *division* problem.

$W = P \div \%$
$W = 270 \div 15\% =$ **$1,800**

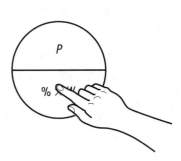

....

In each problem, use the percent circle to determine what it is you're asked to find and circle P, %, or W in the percent circle. Then place a check to indicate whether the problem is solved by multiplication or division.

1. Jimi had to pay a $0.42 sales tax when he bought lunch for $7.00. What tax rate did Jimi pay?

    _____ multiplication   _____ division

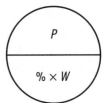

2. Each month, Arnie's net pay is $1,845 and he saves 9% of it. How much does Arnie save each month?

    _____ multiplication   _____ division

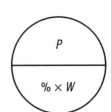

3. To pass her math test, Francine needs to answer 42 questions correctly. If 42 questions is 70% of the test, how many questions are on the test?

    _____ multiplication   _____ division

Finding Part of a Whole

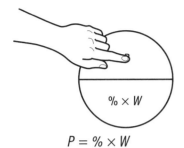

$P = \% \times W$

> **How to Find the Part**
> To find part of a whole, multiply the whole by the percent.

Here are the steps for using a calculator to find part of a whole.

STEP 1 Enter the number representing the whole.

STEP 2 Press the ☒ key.

STEP 3 Enter the percent number.

STEP 4 Press the ☒ key. On most calculators, pressing ☒ completes the calculation. On some calculators, you need to press ☒ and then ☒.

Work through the following examples on your calculator to see if the ☒ key or the ☒ finishes the calculation.

EXAMPLE 1 Find 25% of 70.

Using a Calculator	**Compare with Pencil-and-Paper Solution**
STEP 1 Identify % and *W*. % = 25%, *W* = 70	**STEP 1** Identify % and *W*. % = 25%, *W* = 70
STEP 2 Multiply 70 × 25% on your calculator.	**STEP 2** Change 25% to a decimal by moving the decimal point two places to the left. 25% = 0.25

Press Keys	Display Reads
7 0	7 0 .
×	7 0 .
2 5	2 5 .
% *	1 7 . 5

STEP 3 Multiply 70 by 0.25

$$\begin{array}{r} 70 \\ \times\ 0.25 \\ \hline 3\ 50 \\ 14\ 0 \\ \hline 17.50 = \mathbf{17.5} \end{array}$$

ANSWER: 17.5

*On some calculators, you may need to press ☒ to complete the calculation.

> **Discovery:** Try pressing keys in this order: 2 5 % × 7 0 =
> What happens? Some four-function calculators display "0"! If that is so for your calculator, be sure to enter whole numbers first when you solve percent problems.

<u>EXAMPLE 2</u> At a rate of 5%, how much sales tax will be charged on a
$14.80 purchase?

	Press Keys	Display Reads
STEP 1 Identify % and *W*. % = 5%, *W* = $14.80	1 4 . 8 0	14.80
STEP 2 To solve on your calculator, multiply $14.80 by 5%.	×	14.80
	5	5.
	% *	0.74

ANSWER: $0.74

....

Use your calculator to find each number indicated below. The correct keying order is shown for the first problem in each row.

Percents Between 1% and 100%

1. 20% of 90   35% of 240   8% of $25.50   75% of 88

    9 0 × 2 0 % *

2. 92% of 400   12% of $345   50% of 548   4% of $35.50

    4 0 0 × 9 2 % *

Decimal Percents

3. 0.5% of $4   0.7% of 85   5.5% of $30   3.4% of 62

    4 × . 5 % *

Percents Greater than 100%

4. 250% of 18   300% of $58   150% of 90   200% of $56

    1 8 × 2 5 0 % *

5. Each week Kyle saves 15% of the $243.60 paycheck from his part-time job. What amount is Kyle able to save each week?

6. When his $460 property tax bill goes up by 2.5%, how much additional property tax will Kiwon have to pay?

7. If she makes a commission of 3.5% on each sale, how much commission will Sarah earn for selling a $1,598 couch?

*On some calculators, you may need to press = to complete the calculation.

Finding the Percent

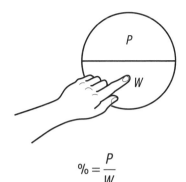

$$\% = \frac{P}{W}$$

How to Find the Percent
To find what percent a part is of a whole, divide the part by the whole.

Here are the steps for using a calculator to find the percent.

STEP 1 Enter the number representing the part.

STEP 2 Press the ÷ key.

STEP 3 Enter the whole.

STEP 4 Press the % key. On most calculators, pressing %
completes the calculation. On some calculators, you need
to press % and then = to finish the calculation.

As the final key pressed, the % key tells the calculator to do the
division and to express the answer as a percent.

EXAMPLE 1 12 is what percent of 60?

Using a Calculator

STEP 1 Identify *P* and *W*.
$P = 12$, $W = 60$

STEP 2 Divide 12 by 60, pressing
% to tell the calculator to
write the answer as a percent.

Press Keys	Display Reads
1 2	12.
÷	12.
6 0	60.
% *	20.

ANSWER: 20%

Compare with Pencil-and-Paper Solution

STEP 1 Identify *P* and *W*.
$P = 12$, $W = 60$

STEP 2 Divide 12 by 60.

$$60\overline{)12.0}^{\,0.2}$$

STEP 3 Change 0.2 to a percent by moving
the decimal point two places to the
right and adding a percent sign.

$$0.2 = 0.20 = \mathbf{20\%}$$

ANSWER: 20%

Discovery: Although your calculator has a % key, no % symbol appears on the display.

EXAMPLE 2 Change $\frac{3}{4}$ to an equivalent percent.

To change $\frac{3}{4}$ to a percent, you are asking, "3 is what percent of 4?" So 3 is the part and 4 is the whole. Looking at the percent circle, you see to divide the part (3) by the whole (4).

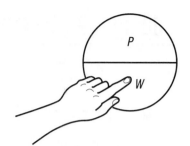

STEP 1 Identify *P* and *W*.
$P = 3$, $W = 4$

STEP 2 To solve on your calculator, divide 3 by 4 and press %.*

Press Keys	Display Reads
3	3.
÷	3.
4	4.
%	75.

ANSWER: 75%

Use your calculator to find each percent. The correct keying order is shown for selected problems.

1. 8 is what percent of 40?

   8 ÷ 4 0 %*

2. 20 is what percent of 25?

3. What percent of 50 is 19?

   1 9 ÷ 5 0 %*

4. What percent of 95 is 38?

5. To the nearest percent, 9 is what percent of 28?

6. To the nearest tenth of a percent, what percent of 145 is 75?

Change each fraction below to an equivalent percent.

7. $\frac{1}{4} =$ $\frac{3}{5} =$ $\frac{9}{20} =$ $\frac{7}{10} =$ $\frac{14}{35} =$

   1 ÷ 4 %*

Change each fraction below to an equivalent percent. If necessary, round your answers to the nearest tenth of a percent.

8. $\frac{3}{8} =$ $\frac{5}{16} =$ $\frac{1}{3} =$ $\frac{3}{32} =$ $\frac{2}{3} =$

9. Out of each $300 paycheck, José's employer withholds $48 for federal income tax. What percent of José's check is withheld for this tax?

10. After riding 10 miles, Louise had completed $\frac{2}{5}$ of the bicycle race. At this point, what percent of the race had Louise completed?

*On some calculators, you may need to press = to complete the calculation.

Finding the Whole

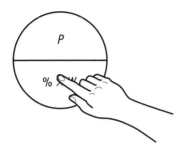

How to Find the Whole
To find a whole when you know the part and the percent, divide the part by the percent.

When using a calculator to find the whole, follow these steps.

STEP 1 Enter the number representing the part.

STEP 2 Press the $\div$ key.

STEP 3 Enter the number of the percent.

STEP 4 Press the $\boxed{\%}$ key. On most calculators, pressing $\boxed{\%}$ completes the calculation. On some calculators, you need to press the $\boxed{=}$ to complete the calculation.

When the percent is less than 100%, then the whole will be *greater* than the part. However, if the percent is greater than 100%, then the whole will be *smaller* than the part.

Discovery: You may notice that the order of keying used to find the *whole* is the same as that used to find the *percent* (discussed on page 77). Be aware of this—but not confused by it! Think of it as simplifying your work.

EXAMPLE 1 16 is 20% of what number? (Or, asked another way, 20% of what number is 16?)

Using a Calculator

STEP 1 Identify P and %.
$P = 16$, % = 20%

STEP 2 Divide 16 by 20% on your calculator.

Press Keys	Display Reads
$\boxed{1}$ $\boxed{6}$	16.
$\boxed{\div}$	16.
$\boxed{2}$ $\boxed{0}$	20.
$\boxed{\%}$	80.

ANSWER: 80

Compare with Pencil-and-Paper Solution

STEP 1 Identify P and %.
$P = 16$, % = 20%

STEP 2 Change 20% to a decimal by moving the decimal point two places to the left.

20% = 0.20

STEP 3 Divide 16 by 0.20

$$0.20\overline{)16.00} = 80.$$

ANSWER: 80

EXAMPLE 2   The sale price of a TV is $289. If $289 is 80% of the original price, what did the TV cost before the sale?

		Press Keys	Display Reads
STEP 1	Identify P and %. P = $289, % = 80%	2 8 9	289.
		÷	289.
STEP 2	To solve on your calculator, divide 289 by 80%.	8 0	80.
		% *	361.25

ANSWER: $361.25

....

Use your calculator to solve each problem below. The correct keying order is shown for problems 2 and 4.

1. 25% of what number is 67?

2. 5.7% of what number is 57?
   5 7 ÷ 5 . 7 % *

3. $14 is 50% of what amount?

4. 52.5 is 35% of what number?
   5 2 . 5 ÷ 3 5 % *

5. 12% of what number is 30?

6. 7.5% of what amount is $150?

7. $32 is 80% of what amount?

8. $90 is 3.6% of what amount?

9. During May, Raymond lost 6 pounds, which is 30% of the weight he hopes to lose on his diet. How many pounds does Raymond hope to lose in all?

10. For the month of January, Guy paid $9.90 in interest charges on his Visa card. Guy pays 1.5% interest charges each month on any unpaid balance. Given these figures, determine the amount of Guy's unpaid balance for the month of January.

11. Darla bought a toaster at the sale shown at the right. Before the sale, what was the price of this toaster?

TOASTER SALE

You pay only
70%
of original price

Sale Price
$12.60

*On some calculators, you may need to press = to complete the calculation.

Increasing or Decreasing a Whole by a Part

Many percent problems involve increasing or decreasing a whole by a part. For example, suppose you want to find the purchase price of a sweater in a state where there is a 5% sales tax. Without a calculator, this can take two steps.

- First, you find the amount of the sales tax (the part).

- Second, you add the sales tax to the original selling price (the whole).

As the example below shows, the % key on your calculator combines these two steps into a single step. This single step is much easier and faster.

EXAMPLE 1   In a state with a 5% sales tax, what is the purchase price of a sweater vest selling for $29.60?

Using a Calculator

STEP 1   Identify % and *W*.
% = 5%, *W* = $29.60

STEP 2   On your calculator, you add $29.60 and 5% of $29.60. The complete calculation is done by pressing the keys shown below!

Press Keys	Display Reads
2 9 . 6 0	29.60
+	29.6
5	5.
% *	31.08

Compare with Pencil-and-Paper Solution

STEP 1   Identify % and *W*.
% = 5%
W = $29.60

STEP 2   Change 5% to a decimal.
5% = 0.05

STEP 3   To find the sales tax, multiply $29.60 by 0.05.

$29.60
× 0.05
1.48 00 = $1.48

STEP 4   To find the purchase price, add the sales tax ($1.48) to the selling price ($29.60).

$29.60
+ 1.48
$31.08

ANSWER: $31.08

ANSWER: $31.08

*On some calculators, you may need to press = to complete the calculation.

Discovery: When you press 2 9 . 6 0 + 5 % *, your calculator automatically **adds** 29.60 and 5% of 29.60. The number 1.48 (5% of 29.60) never appears on the display.

To **subtract** 5%, press 2 9 . 6 0 − 5 % *. Your calculator will automatically subtract 1.48 (5% of 29.60) from 29.60.

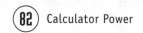

 Below are several types of problems in which you can use your calculator to increase or decrease a whole by a part. Remember:

- **To solve an increase problem, add the percent.**
- **To solve a decrease problem, subtract the percent.**

RATE INCREASE: New amount = original amount + amount of increase

1. Benji's Market sold oat bran for 90¢ per pound. Then he raised the price by 10%. What was Benji's new price per pound for oat bran?

2. In 1980 the population of North Oswego was 48,600. By 1990 the population had risen 24%. In 1990, what was the population?

RATE DECREASE: New amount = original amount – amount of decrease

3. Van had a part-time job that paid her $13,432 per year. She changed jobs, and her new salary is 8.5% lower than her old salary. Find her yearly income at her new part-time job.

4. Due to an unusually warm winter, the Thurmans' heating bill was 30% lower this year than last year. If last year's heating bill was $184.60, how much did they pay this year?

MARKUP: Selling price = store's cost + markup

5. Lucky Saver Stores places a 15% markup on every item they sell. Knowing this, determine the price Lucky Saver will charge for a pair of work boots that cost Lucky Saver $54.60.

6. At Fran's Clothes Closet, Fran pays $56.00 for the Paris Nights evening gowns that she sells. If Fran adds a 30% markup to her cost, what price does she ask for these gowns?

DISCOUNT: Sale price = original price – amount of discount

7. At a Saturday Only sale, Special Electronics is offering a discount of 25% on all store merchandise. At this sale, what will be the price of a car stereo that normally sells for $149.80?

8. The Car Place offers a 10% discount on a case (12 quarts) of car oil. If the regular price is $2.18 per quart, or $26.16 per case, what is the discount price on a full case? Express your answer to the nearest cent.

Percent Problems

Refer to this percent circle as you solve the problems below. In your future work with percents, you may want to draw yourself a percent circle.

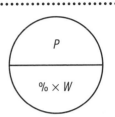

 Problems 1–4 are designed to improve your estimation skills, so use only common sense to answer these questions. For each problem below, circle the choice that best describes the correct answer.

1. Forty-two families send children to Little Cubs Preschool. If 100% of these families attended the school's Fourth of July party, how many families attended in all?

   a. fewer than 42   b. exactly 42   c. more than 42

2. Of the 24 children in Mrs. Grey's fourth grade class, 18 are able to swim. What percent of Mrs. Grey's students can swim?

   a. less than 100%   b. exactly 100%   c. more than 100%

3. Shelley has a part-time job. In June she is getting a 7% raise. What will Shelley's monthly salary be in June if now, in April, she is making $860 per month?

   a. less than $860   b. exactly $860   c. more than $860

4. For 14 of the 34 workers at Emerald Tree Farm, their first language is not English. What percent of the work force do these 14 workers represent?

   a. less than 100%   b. exactly 100%   c. more than 100%

Use your calculator to solve each problem below. As a first step, decide if you are looking for the part, the percent, the whole, or an increase or decrease. Then use the percent circle above to decide whether you should add, subtract, multiply, or divide.

5. When he bought his new dishwasher, Freddie paid a 15% down payment cost of $48. What price did Freddie agree to pay for this appliance?

6. After new speed limits went into effect this year, traffic accidents decreased by 20% each month. How many traffic accidents occurred this July if 75 accidents occurred last July?

7. After paying his bills each month, Bernie has $426 left out of a monthly paycheck of $2,154. To the nearest percent, what percent of Bernie's paycheck does not go to pay bills?

8. At a Memorial Day sale, Glenda saw a coat marked with two tags. One read, "Price $84.00." The other read, "Take an additional 20% off the marked price." If she buys the coat, what can Glenda expect to pay?

9. When she bought a blouse for $28.80, Leona was told that she was paying only 80% of the original price. If this is true, what was the price of the blouse before it was marked down?

10. In an attempt to increase business, Tino has lowered the price of pizzas, as shown at the right. What percent decrease does the price change in the medium size pizzas represent? (**Hint:** Percent decrease = amount of decrease divided by original amount.)

Tino's Pizza Price Reduction!

	Old Price	New Price
Large	$14.75	$13.25
Medium	$12.50	$11.25
Small	$9.85	$8.45

Problems 11–12 refer to the graph below.

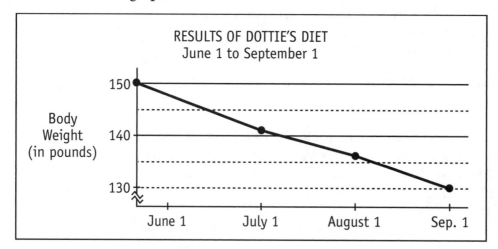

RESULTS OF DOTTIE'S DIET
June 1 to September 1

11. To the nearest percent, what percent of Dottie's June 1st body weight did she lose between June 1st and September 1st?

12. On January 1st, Dottie's weight was 4% less than her June 1st weight. Determine Dottie's weight on January 1st.

Application: Percent and a Family Budget

 To keep track of expenses, many families prepare a household budget. Using a calculator makes it much easier to check and recheck numbers.

Last year the Corwin family had a take-home income of $44,624. At the end of the year, they prepared a record of their yearly expenses. Using this budget, the Corwins are better able to decide how to save money during the coming year.

The two circle graphs below show the Corwins' yearly expenses broken down into various expense categories.

- The graph at the left shows the *dollar amounts* spent.

- The graph at the right (partially completed) shows the *percent* of take-home income spent in each category.

The example shows how the percent for housing (32%) is calculated.

EXAMPLE What percent of their take-home income do the Corwins spend on housing?

> **STEP 1** On the graph at the left, find the dollar amount for housing. $14,280

> **STEP 2** Calculate what percent $14,280 is of $44,624.

Press Keys	Display Reads
1 4 2 8 0 ÷ 4 4 6 2 4 % *	32.000717

ANSWER: 32%

Take-home Pay: $44,624

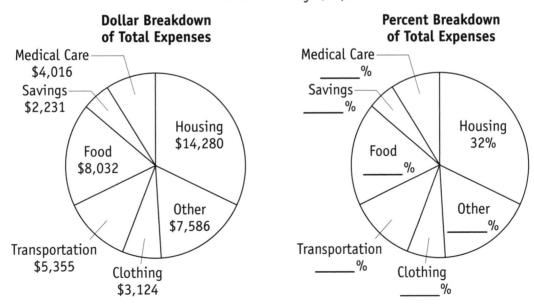

Dollar Breakdown of Total Expenses

Medical Care $4,016
Savings $2,231
Food $8,032
Housing $14,280
Other $7,586
Transportation $5,355
Clothing $3,124

Percent Breakdown of Total Expenses

Medical Care ____%
Savings ____%
Food ____%
Housing 32%
Other ____%
Transportation ____%
Clothing ____%

*On some calculators, you may need to press = to complete the calculation.

Use your calculator and the information on page 85 to do each problem below. Remember to use the percent circle (page 73) to help you decide whether to multiply or divide.

1. Fill in the *Percent Breakdown of Total Expenses* graph on page 85. The percent value for housing is completed as an example. Express each answer to the nearest percent.

2. Last year Mrs. Corwin worked part-time and earned 16% of the family take-home income. What dollar amount of take-home income did Mrs. Corwin earn?

3. Each month last year, the Corwins made a $194.50 car payment. To the nearest percent, what percent of their yearly transportation costs was spent on car payments? (**Hint:** Don't forget to multiply their monthly payment by the number of months in a year.)

4. The Corwins made monthly rent payments of $870.25 during last year. To the nearest percent, what percent of their total take-home pay was spent last year for rent?

5. If the Corwins spent $878.45 last year on prescription drugs, what percent of their medical care expenses was spent for these medicines? Express your answer to the nearest percent.

6. Last year the Corwins donated 1.25% of their total take-home income to charities. To the nearest dollar, what amount did they give to charities?

7. To save money this year, the Corwins have decided to cut back on transportation costs. By taking the bus to work, Mr. Corwin figures he can save 15% on family transportation expenses. If he's right, how much money can Mr. Corwin actually save by riding the bus?

8. The Corwin family has set a goal of putting 25% more money into savings this year than they did last year. To achieve this goal, how much will they need to place in savings this year?

Brain Teasers

You've now reached the end of the chapter on percent! If you like challenges, try the brain teasers on this page. They'll sharpen your thinking skills about calculators and percent problems.

Match each question with the calculator solution that's used to answer that question.

Question	Calculator Solution
_____ 1. What is 18 + 15% of 18?	**a.** [1] [5] [÷] [1] [8] [%]
_____ 2. What percent of 18 is 15?	**b.** [1] [8] [×] [1] [5] [%]
_____ 3. 15% of what number is 18?	**c.** [1] [8] [÷] [1] [5] [%]
_____ 4. What is 18% of 15?	**d.** [1] [8] [+] [1] [5] [%]
_____ 5. What is 15% of 18?	**e.** [1] [5] [×] [1] [8] [%]

 See if you can guess what answer the calculator will display if you press each group of keys as shown below. After you guess, use your calculator to find the actual answer to each problem.*

		Guess	Actual
6.	[1] [0] [0] [×] [1] [0] [0] [%]	_____	_____
7.	[1] [0] [0] [×] [5] [0] [%]	_____	_____
8.	[1] [0] [×] [5] [0] [%]	_____	_____
9.	[1] [0] [0] [÷] [1] [0] [0] [%]	_____	_____
10.	[1] [0] [0] [÷] [5] [0] [%]	_____	_____
11.	[1] [0] [÷] [5] [0] [%]	_____	_____
12.	[1] [0] [0] [+] [1] [0] [0] [%]	_____	_____
13.	[1] [0] [0] [+] [5] [0] [%]	_____	_____
14.	[1] [0] [+] [5] [0] [%]	_____	_____

*On some calculators, you may need to press [=] to complete the calculation.

USING THE CALCULATOR'S MEMORY

You can use the **calculator's memory** to store a number. Unless you erase this number or turn off your calculator, the stored number will remain in memory.

A calculator memory is **cumulative.** This means you can add to or subtract from a stored number as many times as you wish.

The calculator's memory can help you solve problems that involve more than one step.

Exploring Your Calculator's Memory

The Memory Keys

Most four-function calculators have the following memory keys:

M+ , M− , MR , MC

Uses of Memory Keys

M+ and M− On most calculators, pressing either M+ or M− completes a calculation and displays the answer. (On some calculators, you must press = to complete the calculation.)

- Pressing M+ adds the displayed number to the memory.
- Pressing M− subtracts the displayed number from the memory.

MR Memory Recall: Pressing MR displays the total currently stored in memory.

MC Memory Clear: Pressing MC clears (erases) the memory but not the display. (Some calculators have a single MR/MC key. On these calculators, press the MR/MC key once to recall memory and twice to clear the memory.)

(Your calculator may not have these exact memory functions. Also, the M may have a different location on the display. If your keyboard is different, read your calculator's instruction booklet.)

EXAMPLE 1 Add 15 and 12. Put the answer in the memory.

	Press Keys	Display Reads
In the sequence shown at the right, pressing M+ does three things.	1 5	15.
	+	15.
• It completes the addition of 15 + 12.	1 2	12.
• It displays the answer 27.		
• It stores the number 27 in the memory.	M+ *	M 27.
		number in memory ↗
Pressing C (or other clear key) clears the display but not the memory.	C	M 0.
Pressing MR displays the total (27) that is stored in memory.	MR	M 27.
Pressing MC clears the memory, but not the display. The M disappears.	MC	27.

In Example 1, you saw how M+ can be used to complete a calculation and to store the answer in memory. In Example 2, you will see how a calculation can be done within the memory itself.

> **Discovery:** When a calculator has a value stored in memory, a small *M* or the word *memory* appears.

*On some calculators, you may need to press = to complete the calculation before pressing M+ .

Always clear *both* the memory *and* the display before starting a new problem. Before beginning Example 2,

- press MC once or MR/MC twice to clear the memory
- press your clear key to clear the display

EXAMPLE 2 Without using the − key or + key, subtract $36.89 and $19.95 from $87.75.

		Press Keys	Display Reads
STEP 1	Enter 87.75 on the display. Press M+ to place 87.75 in the memory.	8 7 . 7 5 / M+	87.75 / 87.75 M
STEP 2	Enter 36.89 on the display. Press M− to subtract 36.89 from the number (87.75) already stored in memory.	3 6 . 8 9 / M−	36.89 M / 36.89 M
STEP 3	Enter 19.95 on the display. Press M− to subtract 19.95 from the number (87.75 − 36.89, or 50.86) that is currently stored in the memory.	1 9 . 9 5 / M−	19.95 M / 19.95 M
STEP 4	Press MR to display the new total (87.75 − 36.89 − 19.95) that is now stored in the memory.	MR	30.91 M

ANSWER: $30.91

Discovery: After pressing M+ or M−, you do not need to clear the display. The display clears automatically as you begin to enter the next number.

Both Example 1 and Example 2 can easily be done without the use of memory keys. However, working with familiar problems here will help you gain confidence in the use of your calculator's memory.

Match each key with the function it performs.

Keys	Key Functions
_____ 1. M+	a. Displays the total that is currently in memory.
_____ 2. M−	b. Clears the memory but not the display.
_____ 3. MR	c. Clears the display but not the memory.
_____ 4. MC	d. Subtracts the displayed number from the memory.
_____ 5. C	e. Adds the displayed number to the memory.

Arithmetic Expressions

Memory keys can simplify many types of multistep word problems. Here is an example.

EXAMPLE   Ellen is buying six bottles of hair conditioner. Each bottle costs $2.89. How much change will Ellen receive if she pays with a $20 bill?

Instead of solving this example in two steps, you can write the solution as a single expression, called an **arithmetic expression.**

Change = $20.00 −($2.89 × 6)

total cost of
conditioner

- An arithmetic expression consists of numbers, products, and quotients combined by plus and minus signs.

- You can use an arithmetic expression to write the solution steps for any multistep word problem.

You'll soon see how your calculator can help you quickly find the value of arithmetic expressions. First, though, it's a good idea to make sure you understand how to interpret arithmetic expressions.

Evaluating Arithmetic Expressions

- Find the value of numbers within parentheses before adding or subtracting numbers standing alone.

Arithmetic Expressions	Meaning
(19 + 11) − 7	Add 19 and 11; then subtract 7.
(15 × 6) − 13	Multiply 15 × 6; then subtract 13.
(125 ÷ 9) + 14	Divide 125 by 9; then add 14.
75 + (13 × 9)	Multiply 13 × 9; then add 75.
93 − (48 ÷ 12)	Divide 48 by 12; then subtract this quotient from 93.
(9 × 7) + (11 × 6)	Find the products of 9 × 7 and 11 × 6. Then add the products.
(8 × 5) − (4 × 3)	Find the products of 8 × 5 and 4 × 3. Then subtract.

- A number in front of the parentheses indicates multiplication.

Arithmetic Expressions	Meaning
7(8 + 4)	Add 8 and 4; then multiply the sum by 7.
9(6 − 3)	Subtract 3 from 6; then multiply the difference by 9.
(16 − 7) ÷ 5	Subtract 7 from 16; then divide this difference by 5.

Complete the meaning of each arithmetic expression.

Arithmetic Expressions   **Meaning**

1. $(26 + 14) - 9$ Add __26__ and __14__; then subtract __9__.

2. $(13 \times 4) - 18$ Multiply _____ × _____; then subtract _____.

3. $(64 \div 8) + 11$ Divide _____ by _____; then add _____.

4. $21 + (9 \times 8)$ Multiply _____ × _____; then add _____.

5. $60 - (56 \div 7)$ Divide _____ by _____; then subtract this quotient from _____.

6. $(7 \times 6) + (8 \times 3)$ Add the product of _____ × _____ to the product of _____ × _____.

7. $(9 \times 8) - (7 \times 6)$ Subtract the product of _____ × _____ from the product of _____ × _____.

8. $9(7 + 6)$ Add _____ and _____; then multiply the sum by _____.

9. $8(12 - 10)$ Subtract _____ from _____; then multiply the difference by _____.

10. $(30 - 12) \div 6$ Subtract _____ from _____; then divide the difference by _____.

Using a Calculator to Evaluate Expressions

Expressions Involving Only Addition and Subtraction

When only addition and subtraction are involved, press keys in the order of the steps given.

EXAMPLE 1 Find the value of $57 + 19 - 26$.

You do not need to use memory keys to solve this type of problem.

Press Keys	Display Reads
5 7	5 7 .
+	5 7 .
1 9	1 9 .
−	7 6 .
2 6	2 6 .
=	5 0 .

ANSWER: 50

Expressions Involving Sums or Differences in Parentheses

When an operation appears in parentheses, evaluate that sum or difference first. Then perform the rest of the operations.

EXAMPLE 2 Find the value of $15(23 - 7)$.

As shown, first find the difference $23 - 7$. Then multiply the difference by 15.

Pressing the keys in this order ensures that you subtract before you multiply.

Press Keys	Display Reads
2 3	2 3 .
−	2 3 .
7	7 .
×	1 6 .
1 5	1 5 .
=	2 4 0 .

ANSWER: 240

Evaluate the next two arithmetic expressions on your calculator. Memory keys are not needed for these problems.

1. Evaluate $25(56 - 29)$.

   Press keys: 5 6 − 2 9 × 2 5 =

2. Evaluate $(127 - 83) \div 11$.

   Press keys: [1] [2] [7] [−] [8] [3] [÷] [1] [1] [=]

....

Expressions Involving Separated Terms

Terms are numbers, products, and quotients. **Separated terms** have plus (+) or minus (−) signs between them. Use the [M+] key to add a term to memory. Use the [M−] key to subtract a term from memory.

EXAMPLE 3 Evaluate $275 - (13 \times 9)$.

		Press Keys	Display Reads
STEP 1	Place the whole number 275 in the memory.	[2] [7] [5]	275.
STEP 2	Multiply 13×9 and subtract the product from memory.	[M+]	275.^{M}
		[1] [3]	13^M
Note:	Pressing [M−] completes the calculation **and** subtracts the product from the memory.	[×]	13^M
		[9] *	9^M
STEP 3	Press [MR] to display the answer— the total now stored in memory.	[M−]	117^M
		[MR]	158^M

ANSWER: 158

....

Evaluate the next two arithmetic expressions on your calculator. Be sure to clear the memory *and* the display between each problem.

3. Evaluate $(35 \div 8) + 16.75$.

   Press keys: [3] [5] [÷] [8] * [M+] [1] [6] [·] [7] [5] [M+] [MR]

4. Evaluate $(24 \times 19) + (13 \times 12)$.

   Press keys: [2] [4] [×] [1] [9] * [M+] [1] [3] [×] [1] [2] * [M+] [MR]

*On some calculators, you may need to press [=] before [M+] or [M−].

Gaining Confidence with Arithmetic Expressions

On the previous two pages, we've shown how a calculator can be used to find the value of arithmetic expressions. The calculator exercises on this page and the next are designed to help you gain confidence in this newly learned skill.

 Find the value of each arithmetic expression. The correct keying is shown for the first problem in each group.

Expressions Involving Only Addition and Subtraction

Press keys in order of steps given. Memory keys are not used.

1. $28 - 17 + 14 =$
   [2] [8] [−] [1] [7] [+] [1] [4] [=]

2. $\$12.43 + \$9.36 - \$4.06 =$

3. $396 - 209 - 43 + 37 =$

4. $207 + 111 - 73 - 61 =$

5. $\$13.46 - \$2.09 - \$3.12 =$

6. $\$4.25 - \$2.19 - \$1.18 + \$0.24 =$

Expressions Involving Parentheses

Always perform calculations inside the parentheses first. Then multiply or divide as indicated. Memory keys are not used.

7. $12(43 + 7) =$
   [4] [3] [+] [7] [×] [1] [2] [=]

8. $(56 - 28) \div 4 =$

9. $(\$4.56 + \$2.31) \times 3 =$

10. $(45 + 28 - 19) \div 9 =$

11. $14(63 - 35) =$

12. $(29 - 13) \times 5 =$

13. $(\$6.25 + \$2.85) \div 2 =$

14. $(112 + 109 + 121) \div 3 =$

Expressions Involving Separated Products and Quotients

Use the $\boxed{M+}$ key to add a term to memory and use the $\boxed{M-}$ key to subtract a term from memory.
(**Remember:** Clear the memory and the display as you begin a new problem.)

15. $156 + (18 \times 3) =$

$\boxed{1}\,\boxed{5}\,\boxed{6}\,\boxed{M+}\,\boxed{1}\,\boxed{8}\,\boxed{\times}\,\boxed{3}\,*\,\boxed{M+}\,\boxed{MR}$

16. $9 + (104 \div 13) =$

17. $18.7 + (4.2 \times 3) =$

18. $(7.2 \times 6) + (3.4 \times 7) =$

19. $(84 \times 3) - (57 \times 2) =$

20. $37 - (3 \times 4) =$

21. $15 - (153 \div 17) =$

22. $27.6 - (7.8 \times 2) =$

23. $(\$3.79 \times 4) + (\$2.87 \times 5) =$

24. $\$50.00 - (\$13.49 \times 2) - (\$5.88 \times 3) =$

Mixed Practice

25. $\$7.56 + \$2.89 - \$3.57 =$

26. $5(\$8.29 - \$6.42) =$

27. $(\$2.49 \times 3) + (\$5.18 \times 2) =$

28. $\$10.00 - \$2.99 - \$1.89 =$

29. $3(143 + 231) =$

30. $(5.7 + 3.6) \times 7 =$

31. $(\$27.46 - \$17.32) \div 2 =$

32. $(\$14.68 + \$16.20 + \$12.44) \div 3 =$

33. $(\$5.89 - \$2.99) \times 4 =$

34. $\$25.00 - (\$5.25 \times 3) + (\$4.89 \times 2) =$

*On some calculators, you may need to press $\boxed{=}$ before $\boxed{M+}$ or $\boxed{M-}$.

Multistep Word Problems

Examples 1 and 2 show how to use the full power of your calculator to solve a multistep word problem. The first step in each example is to write the solution as an arithmetic expression.

EXAMPLE 1 Lex brought 196 colored markers to school. He gave 84 markers to students in his morning class. He divided the rest equally among the 16 students in his afternoon class. How many markers did each student in the afternoon class get?

		Press Keys	**Display Reads**
STEP 1	Write the expression. $(196 - 84) \div 16$	[1] [9] [6]	196.
STEP 2	Evaluate the expression as shown at the right.	[−]	196.
		[8] [4]	84.
		[÷]	112.
		[1] [6]	16.
		[=]	7.

ANSWER: 7 markers per student

EXAMPLE 2 Kara bought 3 new T-shirts at a cost of $18.99 each. If she pays with a $100 bill, how much change will she receive?

		Press Keys	**Display Reads**
STEP 1	Write the expression. $100.00 - (\$18.99 \times 3)$	[1] [0] [0] [·] [0] [0]	100.00
STEP 2	Evaluate the expression as shown at the right.	[M+]	100.ᴹ
		[1] [8] [·] [9] [9]	18.99ᴹ
		[×]	18.99ᴹ
		[3]	56.97ᴹ
		[M−]	56.97ᴹ
		[MR]	43.03ᴹ

ANSWER: $43.03

Circle the one arithmetic expression that will give the correct answer to each problem. Then use your calculator to evaluate that expression. Use memory keys if they are helpful.

1. Sean gave the clerk $25.00 to pay for a $14.79 pillow and a $4.99 pillowcase. How much change should Sean receive?

   a. $25.00 − $14.79 + $4.99   **b.** $25.00 − $14.79 − $4.99   **c.** $14.79 + $4.99 − $25.00

2. Bill and two friends agreed to divide the $13.95 rental cost of a VCR and two movies, each movie renting for $1.99. To the nearest cent, what is Bill's share of the total cost?

   a. ($13.95 + $1.99) ÷ 3
   b. ($13.95 − $1.99 − $1.99) ÷ 3
   c. ($13.95 + $1.99 + $1.99) ÷ 3

3. Before writing a check for $41.49 and making an $85.00 deposit, Manuel had a checking balance of $152.90. What is his new balance?

   a. $152.90 − $41.49 − $85.00
   b. $152.90 + $41.49 − $85.00
   c. $152.90 − $41.49 + $85.00

4. Jamie received a tax refund of $337.40. She put $50.00 of it into a savings account and divided the rest into 12 equal amounts, one for each month of the coming year. How much money was in each month's amount?

   a. ($337.40 + $50.00) ÷ 12
   b. ($337.40 − $50.00) ÷ 12
   c. ($337.40 − $50.00) ÷ 13

5. Jeb's truck can haul 11.5 cubic yards of dirt per load. Jeb was able to haul 8 full loads on Saturday and 9 full loads on Sunday. How many cubic yards of dirt did Jeb haul on these two days?

   a. (9 + 8) × 11.5
   b. (9 − 8) × 11.5
   c. (9 + 8) ÷ 11.5

6. Every Monday, Wednesday, and Friday, Jake jogs 5 miles during his morning workout. Each Tuesday and Thursday he jogs 3 miles. How many total miles does Jake jog each week?

   a. (3 + 2) × 8
   b. (3 × 5) + (2 × 3)
   c. (5 + 3) × 5

Write an arithmetic expression that shows how to solve each problem below. Then use your calculator to find the value of each expression you've written. Use memory keys if they are helpful.

7. Before writing a check for $18.75 and making a deposit of $125.94, Judy had $84.29 in her checking account. What is the new balance in her account?

expression: _____

value: _____

8. After writing a check for $51.38 and making a deposit of $53.00, Erik's checking balance is $241.85. Determine Erik's balance *before* these transactions.

expression: _____

value: _____

9. Vince walks to work and back each day, 5 days per week. If he lives 1.9 miles from work, how many miles does Vince walk each week going to and from work?

expression: _____

value: _____

10. How much lighter than 20 pounds is a group of four packages if each package weighs 4.3 pounds?

expression: _____

value: _____

11. Cora bought a new clothes dryer for $329.99. She made a down payment of $44.99 and agreed to pay off the balance in three equal monthly payments. If she pays no interest, how much will Cora pay each month?

expression: _____

value: _____

12. For Christmas, Del bought 4.5 pounds of chicken priced at $3.08 per pound and 6.4 pounds of beef priced at $8.75 per pound. How much more did Del pay for the beef than he paid for the chicken?

expression: _____

value: _____

13. During league play Saturday, Julie bowled games of 182, 175, and 193. To the nearest whole number, what was Julie's average score that evening?

expression: _____

value: _____

Special Multistep Problem: Finding Percent Increase or Decrease

One type of multistep problem involves finding percent increase or percent decrease. These problems combine your percent skills with your skills of evaluating arithmetic expressions.

$$\% = \frac{P}{W} = \frac{\text{part (amount of increase)}}{\text{original amount}}$$

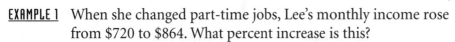

EXAMPLE 1 When she changed part-time jobs, Lee's monthly income rose from $720 to $864. What percent increase is this?

		Press Keys	Display Reads
STEP 1	Subtract to find the **amount of increase**— the part (*P*). $P = \$864 - \$720$	8 6 4	864.
		−	864.
STEP 2	Divide *P* by *W* (the original amount $720) to find the **percent increase** (%).	7 2 0	720.
		÷	144.
		7 2 0	720.
		% *	20.

ANSWER: 20%

Calculate each percent increase or percent decrease. Round your answers to the nearest percent.

$$\% = \frac{P}{W} \qquad \% \text{ increase or decrease} = \frac{\text{amount of increase or decrease}}{\text{original amount}}$$

1. Last month, Cybil's salary was raised from $8.60 per hour to $9.30 per hour. What percent raise did Cybil receive?

2. Nora's Woolens reduced the price of a sweater from $38.50 to $27.95. What percent price reduction is this?

3. Between 1990 and 2000, the value of Will's condominium increased from $48,600 to $62,800. What percent value increase is this?

4. In the past 10 years, the population of Oak Grove decreased from 72,600 to 61,710. What percent population decrease does this change represent?

5. Between September 1 and June 1, the average weekly rainfall in Newton increased from 1.2 inches to 1.5 inches. What percent increase in rainfall does this change represent?

6. For the weekend, Jerry's lowered the price of its single-scoop ice cream cones from $2.25 to $1.75. What percent price decrease is this price reduction?

*On some calculators, you may need to press = to complete the calculation.

The next example shows how to calculate percent increase or decrease using the memory keys.

EXAMPLE 2   Every year, Mary Anne reviews the stocks in her retirement account. One of her technology stocks had a value of $58.75 last year and $73.125 this year. What was the percent increase for the year? (**Note:** Stock prices are often given to the fraction of a cent as in $73.125.)

		Press Keys	Display Reads
STEP 1	Start by storing the old price in memory.	5 8 · 7 5	58.75
STEP 2	Find the amount of increase. Use the value in memory for the old price.	M+	M 58.75
		7 3 · 1 2 5	M 73.125
		−	M 73.125
STEP 3	Divide by the value in memory. (It is still the old price.) Press % to show the answer as a percent.	MR	M 58.75
		÷	M 14.375*
		MR	M 58.75
		% **	M 24.4688085

ANSWER: 24%

If you look at the *Press Keys* column, you can see that only two values were entered in the calculator. All the other steps used single keys on the calculator. Using single keys saves you a lot of time, and lessens the possibilities of keyboarding errors.

• •

Use the memory keys to find the percent increase or decrease for each stock. Write each answer to the nearest tenth of a percent.

Stock	Old Price	New Price		Stock	Old Price	New Price
7. telephone	35.213	45.813	8.	automobile	86.95	88.27
9. bank	22.166	19.588	10.	computer	141.725	165.298
11. transportation	39.07	34.215	12.	pharmacy	86.00	135.28
13. clothing	27.55	36.432	14.	hotel	98.652	91.753
15. publisher	44.14	57.477	16.	travel	102.15	105.972

*The value 14.375 is the difference 73.125 − 58.75. The value 58.75 is still in memory.
**On some calculators, you may need to press = to complete the calculation.

Calculator Posttest A

....

This posttest gives you a chance to check your skills in using a calculator. Take your time and work each problem carefully. When you finish, check your answers and review any topics on which you need more work.

CALCULATOR BASICS

Match each key with its function.

_____ 1. $\boxed{+}$ **a.** recall a stored number

_____ 2. $\boxed{\div}$ **b.** add a number to memory

_____ 3. $\boxed{\text{M+}}$ **c.** add two numbers

_____ 4. $\boxed{\text{MR}}$ **d.** divide two numbers

For problems 5 and 6, show how each number appears in a calculator display.

5. three thousand, five hundred seventy

   ┌─────────────────────────┐
   └─────────────────────────┘

6. seventeen dollars and eight cents

   ┌─────────────────────────┐
   └─────────────────────────┘

For problems 7–15, show the calculator keys you press to solve each problem. You may not need all the keys shown.

7. $2,400 + 750 =$

   □ □ □ □ □ □ □ □ □

8. $200 - 125 - 46 =$

   □ □ □ □ □ □ □ □ □ □ □

9. $\$423.75 \times 6 =$

   □ □ □ □ □ □ □ □ □

10. $36.50 ÷ 5 =

□□□□□□□□

11. Write $\frac{7}{8}$ as a decimal.

□□□□□

In problems 12–15, the final ⬜= **key is not needed on most calculators.**

12. What is 35% of 900?

□□□□□□□□

13. 15 is what percent of 80?

□□□□□□□

14. 24 is 30% of what number?

□□□□□□□

15. Using memory keys, find the value of $(16 \times 5) - (14 \times 3)$.

□□□□□□□□□□□

CALCULATOR APPLICATIONS

 Use your calculator to solve each problem.

16. Terry lost 26 pounds on his diet. If he weighed 214 pounds before starting his diet, what is his weight now?

17. A sale rack at a clothing store has 35 shirts, 17 jackets, and 24 sweaters. How many pieces of clothing are on the rack?

18. A customer buys three shirts from the sale rack for $8.75 each, including tax. If the customer gives the clerk three $10 bills, how much change should she receive?

19. Yogurt is on sale for $0.59 per container. How much do 3 cases of this yogurt cost if each case contains 24 containers?

20. Last week, Keisha worked 38 hours and earned $448.75. For this work, what was her hourly pay rate? Round your answer to the nearest cent.

21. Jan wants to take 170 cookies to share as dessert at the school barbecue. If she can place only 8 cookies on a paper plate, how many paper plates will she need?

22. A machinist wants to make 7 equal spacers. Placed end to end, the 7 spacers will span a distance of 2.5 inches. To the nearest thousandth inch, what should be the width of each spacer?

23. Each month, Danni plans to save 20% of her income. If she earns $1,234 during June, how much should Dannie save?

24. Out of her June earnings of $1,234, Danni's employer withheld $205 for federal income tax. To the nearest percent, what percent of Danni's paycheck is withheld for federal tax?

25. At the annual Spring Fling sale, the Clothes House is offering a 35% discount on all clothes they sell. At this sale, what will be the price of a dress that normally sells for $124?

26. Lam made a down payment of 15% when he bought a new computer. If he paid $210 down, what was the purchase price?

Memory keys may be useful for problems 27–30.

27. Leng bought 4 potted plants at a cost of $8.98 each. If she pays the clerk with a $50 check, how much change should Leng receive?

28. While shopping, Debbie bought 3 pounds of mixed nuts selling at $2.89 per pound and 5 pounds of grapes selling at $1.79 per pound. What total amount did Debbie pay for this purchase?

29. Because of outstanding work, Marie was given a pay raise. Her hourly pay rate increased from $7.50 to $8.25. What percent increase in pay does Marie's raise represent?

30. The price of a new Sony television decreased from $350 to $250. What percent price decrease does this reduction represent? Round your answer to the nearest percent.

Calculator Posttest A Prescriptions

Circle the number of any problem that you miss. A passing score is 26 correct answers. If you passed the test, go on to Using Number Power. If you did not pass the test, review the chapters in this book or refer to these practice pages in other materials from Contemporary Books.

PROBLEM NUMBER	SKILL AREA	PRACTICE PAGES
1, 2, 3, 4, 5, 6	calculator basics	7–12, 89–90
15, 27, 28, 29, 30	using memory keys	89–96
	basic arithmetic:	
	Math Exercises: Whole Numbers & Money	5–27
	Math Skills That Work: Book One	26–163
	Breakthroughs in Math: Book 1	18–133
7, 8, 9, 10, 16, 17, 21	add, subtract, multiply, divide	19–22, 31–36, 40
18, 19	multistep problems	37–38, 97–99
	decimals and fractions:	
	Math Exercises: Decimals	10–11, 25
	Math Skills That Work: Book Two	32–105
	Breakthroughs in Math: Book 2	34–113
11	changing fractions to decimals	66–67
20, 22	rounding decimals	47–48
	percents:	
	Math Exercises: Percents	12–17, 20–22
	Math Skills That Work: Book Two	106–139
	Breakthroughs in Math: Book 2	114–141
12, 23	finding part of a whole	75–76
13, 24	finding a percent	77–78
14, 26	finding the whole	79–80
25	decreasing a whole by a part	81–82

For further basic mathematics practice:

Math Solutions (software)
  Whole Numbers; Percents, Ratios, Proportions

Basic Skills/Pre-GED Interactive (software)
  Mathematics Units 1, 2, 3, 7

GED Interactive (software)
  Mathematics Units 2, 3, 4

Calculator Posttest B

....

This posttest gives you a chance to check your calculator skills in a multiple-choice format as used in the GED test and other standardized tests. Take your time and work each problem carefully before you choose your answer. When you finish, check your answers with the answer key at the back of the book.

Calculator Basics

1. Which key is used to subtract?

   a. M−   b. −   c. MR   d. ×   e. ÷

2. Which key is used to add a number to the calculator's memory?

   a. M−   b. +   c. M+   d. ×   e. ÷

3. Which function or operation is indicated by the ÷ key?

   a. addition   c. subtraction   e. multiplication

   b. division   d. store in memory

4. Which function or operation is indicated by the MR key?

   a. reducing a fraction   c. rounding an answer   e. displaying a remainder

   b. rounding a number stored in memory   d. recalling a number stored in memory

5. Which key does a calculator normally **not** have?

   a. comma key   c. decimal point key   e. division key

   b. add to memory key   d. clear key

6. How would the value "four thousand, six hundred twenty-three" appear on a calculator display?

   a. 4.623   c. 46230   e. 4623.

   b. 46.23   d. 0.4623

7. How would the value "thirty-two dollars and fifty cents" appear on a calculator display?

a. `32.50`   c. `32-50`   e. `$32.50`

b. `0.3250`   d. `32.5`

8. How would the value "seventy-eight thousandths" appear on a calculator display?

a. `78.000`   c. `78000.`   e. `0.78000`

b. `0.780`   d. `0.00078`

In problems 9–15, choose the correct key sequence that solves each problem.

9. Find the sum 350 + 139.

a. [3][5][0][+][1][3][9][%]   c. [3][5][0][+][1][9][3][=]   e. [3][5][0][−][1][3][9][=]

b. [3][0][5][+][1][3][9][=]   d. [3][5][0][+][1][3][9][=]

10. Divide $46.53 into 3 equal amounts.

a. [4][6][.][5][3][÷][3][=]   c. [4][6][5][3][.][÷][3][=]   e. [4][6][.][5][3][=][÷][3]

b. [4][6][5][3][.][3][÷][=]   d. [4][6][÷][5][3][.][3][=]

11. Write the fraction $\frac{3}{4}$ as a decimal.

a. [3][.][4][=]   c. [4][÷][3][=]   e. [3][÷][4][=]

b. [3][−][4][=]   d. [3][×][4][=]

12. Find 18% of 76.

a. [7][6][÷][1][8][%]   c. [7][6][×][1][8][%]   e. [1][8][×][7][6][=]

b. [7][6][.][1][8][%]   d. [7][6][−][1][8][%]

13. What percent of 256 is 32?

    a. | 2 | 5 | 6 | − | 3 | 2 | % | **c.** | 2 | 5 | 6 | + | 3 | 2 | % | **e.** | 2 | 5 | 6 | ÷ | 3 | 2 | % |

    b. | 3 | 2 | ÷ | 2 | 5 | 6 | % | **d.** | 3 | 2 | × | 2 | 5 | 6 | % |

14. 18 is 16% of what number?

    a. | 1 | 8 | × | 1 | 6 | % | **c.** | 1 | 6 | × | 1 | 8 | % | **e.** | 1 | 8 | ÷ | 1 | 6 | % |

    b. | 1 | 6 | ÷ | 1 | 8 | % | **d.** | 1 | 8 | − | 1 | 6 | % |

15. Using memory keys, find the value of $(14 \times 7) + (25 \times 6)$.

    a. | 1 | 4 | 7 | × | 2 | 5 | 6 | M+ | × | M+ | MR |

    b. | 1 | 4 | M+ | 7 | + | 2 | 5 | M+ | × | 6 | MR |

    c. | 1 | 4 | × | 7 | M+ | 2 | 5 | × | 6 | M+ | MR |

    d. | 1 | 4 | × | 7 | M+ | 2 | 5 | × | 6 | M− | MR |

    e. | 1 | 4 | + | 7 | M+ | 2 | 5 | + | 6 | M− | MR |

Applications

Use your calculator to solve each problem.

16. Jonathan paid for a chicken sandwich with $10. If the sandwich cost $3.95, how much change should Jonathan receive?

    a. $6.05   **b.** $2.95   **c.** $4.95   **d.** $6.50   **e.** $13.95

17. For dinner, Elaine and two friends bought pizza for $14.95, drinks for $3.45, and salads for $3.75. What is the total cost of this meal?

    a. $18.70   **b.** $19.35   **c.** $21.65   **d.** $22.15   **e.** $23.85

18. Lateesha bought three T-shirts on sale for $5.79 each. If she pays the clerk $20, how much change should she receive?

    a. $2.63   **b.** $4.21   **c.** $10.43   **d.** $12.45   **e.** $14.21

19. At a canned food sale, Mrs. Landers bought 3 cases of canned peas. Each case contained 24 cans. If each can weighs 12 ounces, how many total ounces of peas did Mrs. Landers buy?

    a. 288   **b.** 582   **c.** 864   **d.** 990   **e.** 1,158

20. Manuel paid $29.40 for 19 gallons of gas. Rounded to the nearest cent, what did Manuel pay per gallon?

    a. $1.49   **b.** $1.55   **c.** $1.59   **d.** $1.62   **e.** $1.67

21. Parents are needed to drive a class of 28 students on a field trip. If each parent can take 5 students, how many parents are needed?

    a. 3   **b.** 4   **c.** 5   **d.** 6   **e.** 7

22. A jeweler divides 43 grams of gold into 6 equal parts. To the nearest thousandth gram, what is the weight of each part?

    a. 7.16   **b.** 7.17   **c.** 7.166   **d.** 7.167   **e.** 7.1667

23. During the last presidential election, only 38% of the registered voters in Lynn County voted. How many of the 26,850 registered voters in Lynn County voted?

    a. 10,203   **b.** 11,150   **c.** 13,425   **d.** 16,850   **e.** 38,000

24. While on a diet, Wendi lost 19 pounds. If she weighed 178 pounds before starting the diet, what percent of her body weight did Wendi lose? Round your answer to the nearest percent.

    a. 8%   **b.** 11%   **c.** 14%   **d.** 19%   **e.** 23%

25. The enrollment at Hoover Elementary School is 8% greater this year than last year. If 425 students attended Hoover last year, how many students are attending this year?

    a. 391   **b.** 425   **c.** 459   **d.** 491   **e.** 559

26. Twenty-nine percent of the customers who shop at Haley's Department Store are men. If 720 men shopped at Haley's last month, about how many total customers did Haley's have during the month? Round your answer to the nearest 100.

    a. 1,500   b. 1,600   c. 1,800   d. 2,100   e. 2,500

Memory keys may be useful for problems 27–30.

27. Sammi is driving to Chicago, a distance of 394 miles. Sammi drove for 3 hours before stopping for lunch. If he averaged 58 miles each hour, how many miles is Sammi from Chicago now?

    a. 220   b. 278   c. 340   d. 392   e. 426

28. At Twin Pines Hardware, Denise is buying 6 gallons of white paint at $12.48 per gallon and 3 quarts of blue paint at $4.29 per quart. How much should Denise be charged in all?

    a. $67.35   b. $72.95   c. $79.45   d. $87.75   e. $93.25

29. During the last year, the average price of a new home in Benton County rose from $125,000 to $137,500. What percent price increase does this change represent?

    a. 9%   b. 10%   c. 11%   d. 13%   e. 14%

30. Joey's Pizza is reducing the price of their Joey's Special from $14.75 to $12.50. To the nearest percent, what percent price decrease does this reduction represent?

    a. 9%   b. 11%   c. 13%   d. 15%   e. 18%

Calculator Posttest B Chart

Circle the number of any problem that you miss and review the appropriate practice pages. A passing score is 26 correct answers. If you passed the test, go on to Using Number Power. If you did not pass the test, take the time to make a more thorough review of the book.

PROBLEM NUMBER	SKILL AREA	PRACTICE PAGES
1, 2, 3, 4, 5, 6, 7, 8	calculator basics	7–12, 89–90
9, 17	adding	19–20
10	dividing	33–34
11	changing fractions to decimals	66–67
12, 23	finding part of a whole	75–76
13, 24	finding a percent	77–78
14, 26	finding the whole	79–80
16	subtracting	21–22
18, 19	multistep problems	37–38, 97–99
20, 22	rounding decimals	47–48
21	division with a remainder	40
25	increasing a whole by a part	81–82
15, 27, 28, 29, 30	using memory keys	89–96

Using
Number
Power

Keeping a Mileage Record

 People who use a car for work keep mileage records. Frank Barclay works as a truck driver for Lewiston Trucking Company. His partially completed weekly mileage record for the week of April 15 is shown below.

LEWISTON TRUCKING COMPANY WEEKLY MILEAGE RECORD

Week Of: April 15   **Name: Frank Barclay**

	Date	Odometer Begins	Odometer Ends	Daily Mileage
Mon	4/15	38,643	39,025	382
Tue	4/16	39,025	39,432	
Wed	4/17	39,432	39,786	
Thu	4/18	39,786	40,216	
Fri	4/19	40,216	40,504	
Sat	4/20	40,504	40,880	
Sun	4/21	40,880	41,409	
			Weekly Total:	

To compute each daily mileage, subtract the number under Odometer Begins from the number under Odometer Ends.

EXAMPLE   Determine the number of miles Frank drove on Monday, April 15.

Odometer Ends   39,025
Odometer Begins   − 38,643
                          382

ANSWER: 382 miles

 Use your calculator to help answer each question below.

1. Fill in the Daily Mileage column in the record above for Tuesday through Sunday.

2. Determine the total miles Frank drove during the week of April 15.

3. How many miles did Frank drive

   a. during the weekend (Saturday and Sunday)?

   b. during the week (Monday through Friday)?

Completing a Purchase Order

A **purchase order** (also called a **supply order**) is a form that a company fills out when it orders products from another company. Tina Vinson works for Downtown Hardware. Part of her job responsibility is to order items for the store as supplies run low. Her partially completed purchase order to Western Industrial Supply is shown below.

WESTERN INDUSTRIAL SUPPLY

	Item #	Description	Quantity	Cost/Per	Total Amount
1.	29-75A	Cordless Drill	12	$32.45	$389.40
2.	34-08V	3-Drawer Tool Chest	5	$39.99	
3.	03-14A	Star Claw Hammer	9	$12.95	
4.	07-24C	Deluxe Square Shovel	17	$18.88	
5.	47-83B	Bench-Grip Vise	8	$15.49	
6.	62-80B	10-Foot Ladder	6	$56.29	
7.	09-12A	#6 Screwdriver Set	16	$9.65	

Purchaser: _Tina Vinson_ | Total Purchase

To compute each Total Amount, multiply the Cost/Per times the Quantity. The Cost/Per is the price for each single described item.

EXAMPLE Determine the amount that Downtown Hardware must pay for 12 Cordless Drills.

cost/drill $32.45
× quantity × 12
total amount $389.40

ANSWER: $389.40

Remember: Your calculator will display $\boxed{389.4}$ for $389.40. It does not show a 0 at the right of a decimal answer.

Use your calculator to help answer each question below.

1. Complete the Total Amount column above for items 2 through 7.

2. Compute the Total Purchase by adding the seven entries in the Total Amount column.

3. When the order arrived, there was a note saying that Western Industrial no longer carried Star Claw Hammers. Since these hammers weren't shipped, what is the new Total Purchase amount?

Interpreting a Paycheck Stub

 An employer gives an employee a paycheck stub to tell the employee how much money is being withheld from his or her paycheck. The stub lists each **deduction** (withheld amount).

LAPINE MANUFACTURING
Employee: Jon Allen

	Gross Pay	Federal Income Tax	State Income Tax	Social Security	Net Pay
Current Pay Period (two weeks)	$684.80	$79.24	$14.27	$51.72	$539.57
Year-to-Date	$8,217.60	$950.88	$171.24	$620.64	$6,474.84

Answer the questions below by filling in each blank with a number taken from the paycheck stub.

1. **Gross pay** is the total amount of money earned each pay period.
    Net pay is the actual amount of a paycheck.
    net pay = gross pay – total deductions
    What two ways can you use to determine Jon's total deductions for this pay period?
    a. Subtract _____ from _____, or
    b. Add _____, _____, and _____

2. Each **year-to-date** amount is the total sum of that amount including each pay period up through the present date.
    To determine how many pay periods Jon has been paid this year, divide _____ by $684.80.

3. To determine how much total federal income tax will be withheld from Jon's check during an entire year, multiply _____ by 26 (the number of pay periods in one year).

 Use your calculator to help answer the following questions.

4. How much are Jon's total deductions for the pay period shown?

5. How many pay periods this year has Jon worked for Lapine?

6. If Jon works 42.8 hours each pay period, find to the nearest cent

    a. Jon's gross pay per hour   **b.** Jon's net pay per hour

7. About how much does Jon pay each year for social security?

Comparing Annual Car Costs

For their new business, Sherm and Melissa Harland have decided to buy one of two used pickup trucks. Their decision will be based on a comparison of first-year costs of each truck.

Use your calculator to complete the lists of first-year costs of each truck described below. Fill in the correct amount on each blank line.

Immediate Costs	CHEVROLET	FORD
Purchase price	$8,645.00	$7,975.00
License & registration	132.50	132.50
Tune-up	149.95	164.95
New tires	329.88	none
Radiator repair	178.99	none
Brake repair	none	238.49
New muffler	129.99	135.99
1. Insurance premium (every three months)	$225.00 _____ (yearly)	$248.00 _____ (yearly)
2. **Total Immediate Costs:**	_____	_____

Operating Expenses		
Gas Expense:		
Estimated yearly mileage	12,000	12,000
Miles per gallon	17	14
3. Number of gallons needed (to nearest gallon)	_____	_____
Cost per gallon	$1.47	$1.47
4. Total cost of gas (to nearest cent)	_____	_____
Oil Expense:		
5. Three oil changes at $24.95 each	_____	_____
6. **Total Operating Expenses:**	_____	_____
7. **Total First-Year Costs:** (Add line 2 and line 6.)	_____	_____

Completing Payroll Forms

 Calculators are used in many workplaces, especially for repetitive tasks. A repetitive calculator task is one in which you make similar calculations over and over again.

A calculator really simplifies the work of filling out a company's payroll forms. Each employee's hours must be added, and both total pay (**gross pay**) and take-home pay (**net pay**) must be calculated.

In the problems below, use your calculator to fill out the weekly payroll forms of Harding Tool Company. At Harding, a regular work week is 40 hours. Any hours beyond 40 are paid at an overtime rate of 1.5 times the regular hourly rate (commonly called "time and a half").

For fun, time yourself on these two exercises. When you finish, estimate how long it would take you to do a 50-person payroll!

 1. **Fill in the Determining Hours Worked chart. As a guide, Laurie Allen's row has been filled in for you.**

EXAMPLE   Hours for Laurie Allen

STEP 1		STEP 2	STEP 3	
Add	8.0	Regular hours: 40	Subtract	
	8.0	All the hours	Total hours:	45.0
	9.0	worked up to	Regular hours:	− 40.0
	9.5	40 hours for five days are	Overtime:	5.0
	+ 7.5	regular hours.		
	+ 3.0			
	45.0			

Discovery: As you work these exercises, keep your eye on the display to avoid simple keying errors. Be sure you enter the numbers you intend to.

DETERMINING HOURS WORKED

Name	M	T	W	T	F	S	S	①Total Hours	②Regular Hours	③Overtime Hours
Allen	8.0	8.0	9.0	9.5	7.5	3.0		45	40	5
Cook	8.0	9.5	6.5	9.0	8.5	2.5	2.5		40	
Dart	7.5	9.0	5.5	7.5	6.5				36	
Franks		8.5	7.5	8.5	9.0	8.5	3.5		40	
Norris	9.0	8.5	8.0	5.5	7.5	8.5			40	

2. **Fill in the Determining Net Pay chart. The first step is to compute the net pay for each employee. Again, Laurie Allen's row has been filled in for you as an example.**

EXAMPLE   Net pay for Laurie Allen

STEP 1   Regular hours: <u>40</u>
(from Determining Hours Worked chart, page 118)

STEP 2   Regular pay rate: <u>$8.32</u>
(from Determining Net Pay chart below)

STEP 3   Total regular pay
Multiply
$8.32 × 40 = <u>$332.80</u>

STEP 4   Overtime hours: <u>5</u>
(from Determining Hours Worked chart, page 118)

STEP 5   Overtime pay at time-and-a-half
Multiply
$8.32 × 1.5 = <u>$12.48</u>

STEP 6   Total overtime pay
Multiply
$12.48 × 5 = <u>$62.40</u>

STEP 7   Total pay
Add
$332.80 + $62.40 = <u>$395.20</u>

STEP 8   Withholding: <u>18%</u>
(from Determining Net Pay chart below)

STEP 9   Net Pay
Subtract 18% of $395.20 from $395.20.

| 3 | 9 | 5 | · | 2 | 0 | − | 1 | 8 | % |*

The displayed answer, | 324.064 |, rounds to <u>$324.06.</u>

*On some calculators, you may need to press | = | to complete the calculation.

Find each employee's net pay and complete the chart.

DETERMINING NET PAY

Name	① Regular Hours	② Regular Pay Rate	③ Total Regular Pay	④ Overtime Hours	⑤ Overtime Pay Rate	⑥ Total Overtime Pay	⑦ Total Pay	⑧ Total % Withholding	⑨ Net Pay
Allen	<u>40</u>	$8.32	$332.80	<u>5</u>	$12.48	$62.40	$395.20	18%	$324.06
Cook	___	$8.32	___	___	___	___	___	18%	___
Dart	___	$9.28	___	___	___	___	___	19%	___
Franks	___	$9.48	___	___	___	___	___	19%	___
Norris	___	$10.94	___	___	___	___	___	21%	___

Total Regular Hours For All Workers _____
  (Add column 1.)
Total Overtime Hours For All Workers _____
  (Add column 4.)
Total Regular Pay For All Workers _____
  (Add column 3.)

Total Net Pay For All Workers _____
  (Add column 9.)
Total Overtime Pay For All Workers _____
  (Add column 6.)

Ordering From a Menu

 Every day, millions of people eat at fast food restaurants. The menu below is probably similar to one you've seen many times.

JERRY'S BURGER CITY

Sandwiches/Salads		Drinks	
Jerry's Single	$0.94	Soft Drinks	$0.89, $1.09
Single Cheeseburger	2.09	Milk	0.85, 1.10
Jerry's Double	2.49	Coffee	0.95, 1.20
Double Cheeseburger	2.69	Tea	0.95
Chicken Sandwich	2.89	Hot Chocolate	1.25
Fish Sandwich	2.79		
Ham Sandwich	2.19		
		Desserts	
French Fries	$0.75, 1.25	Pie	$2.39
Dinner Salad	2.49	Cake	2.39
Salad Bar	3.29	Ice Cream	2.00

  Use your calculator to help answer each question. Use estimation to make sure your answer makes sense.

1. Robert is at Jerry's Burger City with his two children. He has only $12.00 to buy lunch for the three of them. Paying $0.85 for tax, can he afford to buy the lunch listed at the right?

> 2 Single Cheeseburgers
> 1 Double Cheeseburger
> 2 Small French Fries
> 2 Small Milks
> 1 Large Coffee

2. After the game Saturday night, Danny and Gary stopped at Jerry's and ordered the food shown at the right. Before taxes, how much will this meal cost them?

> 1 Double Cheeseburger
> 1 Chicken Sandwich
> 2 Large French Fries
> 1 Dinner Salad
> 2 Large Soft Drinks
> 2 Pieces of Pie

3. During her noon lunch break, Phyllis stopped by Jerry's for something to eat. She ordered a ham sandwich, a dinner salad, and a hot chocolate. How much was tax and tip if Phyllis spent a total of $7.38?

"Best Buy" Shopping

 Each time you make a purchase, you are faced with the question, "Which brand shall I buy?" Sometimes you prefer one brand over another, even if it costs more. Other times, you simply want the lowest price.

Unit Price

Unit price is the amount you pay per single unit of purchase.

- If you're buying tomatoes by the pound, the unit price is the price of each pound.

- If you're buying a dozen envelopes, the unit price is the price of each envelope.

When the brand name is not important, you can find the "best buy" by computing the lowest unit price.

- To compute unit price, divide the total price by the number of units purchased.

EXAMPLE 1   Which of the following choices offers the best buy on peaches?

    a. 6 lb for $8.64
    b. 5 lb bag for $6.90
    c. 10 lb bag for $1.41 per pound

To find the price per pound, divide the total price by the number of pounds.

    a. $8.64 ÷ 6 = $1.44   **b.** $6.90 ÷ 5 = $1.38   **c.** Given as $1.41 per lb

ANSWER: Choice **b,** at **$1.38 per pound** for a **5-pound bag,** is the best buy.

 Use your calculator to compute the unit price of each item below. Circle the best buy in each group.

1. orange juice concentrate   **a.** $2.04 for 12 oz   **b.** $3.84 for 24 oz   **c.** $6.08 for 32 oz

2. rolls of film   **a.** 12-exposures for $5.76   **b.** 24-exposures for $11.76   **c.** 36-exposures for $16.92

3. multivitamins   **a.** 75 tablets for $14.25   **b.** 100 tablets for $20.00   **c.** 150 tablets for $31.50

Lowest Total Cost

Total cost is the cost of purchasing several units of an item. To find the total cost multiply unit price × number of items.

For example, to buy three picture frames at $6.98

$$\$6.98 \times 3 = \$20.94$$

For some purchases, though, many stores give a discount when you buy several of the same item. For these purchases, finding the best buy involves taking advantage of discounts.

EXAMPLE 2   When Smell Fresh detergent went on sale at several stores, Patrick decided to buy a case of 12 bottles. Which of the three stores below is offering the best buy?

Freddy's	Home Foods	Value Foods
Smell Fresh	Smell Fresh	Smell Fresh
$1.19 per bottle	$1.29 per bottle	$1.29 per bottle
(no case discount)	$1.50 rebate	$0.15 per bottle discount on
	each case	purchase of 12 or more

To find the best buy, compute the price of 12 bottles at each store.

Freddy's

$1.19 × 12 = $14.28

Home Foods

($1.29 × 12) – $1.50

= $15.49   – $1.50
= $13.98

Value Foods

$1.29 – 0.15 = $1.14
$1.14 × 12 = $13.68

ANSWER: Value Foods is offering the best buy for 12 bottles at $13.68.

 Circle the store in each group that is offering the best buy.

4.   15 pounds of grapes

   Freddy's: $0.98 per pound

   Home Foods: $0.99 per pound,
   $0.50 discount on 15-lb bag

   Value Foods: $1.09 per pound,
   $1.50 rebate for purchases of
   $10.00 or more

5.   12 quarts of skim milk

   Save More: $1.04 per quart

   Buy Right: $1.19 per quart
   $1.00 discount on case of 12

   Marcy's: $1.13 per quart, "Buy
   11 and get 1 free!"

Computing Distance, Rate, and Time

Just think how often you've asked these questions:

How far is it? How fast are we going? How long will it take?

These three quantities—distance, rate, and time—are related by the **distance formula:** *distance* equals *rate* times *time.*

Distance (*d*)	=	**Rate (*r*)**	×	**Time (*t*)**
Usually expressed in miles		Speed—usually expressed in miles per hour (mph)		Usually expressed in hours

Written in short form as $d = rt$, the distance formula is used to find the distance when the rate and time are known. Once again, using a calculator carefully can be a big help.

EXAMPLE 1 The bus between Chicago and New York averages 45 miles per hour. How far can this bus travel in the first 8 hours of the trip?

    STEP 1 Identify r and t.
          $r = 45$ miles per hour $t = 8$ hours

    STEP 2 Substitute values into the distance formula: $d = rt$
          $45 \times 8 =$ **360 miles**

The distance formula can also be expressed as a **rate formula** (the speed) or **time formula.**

rate formula: $r = d \div t$		time formula: $t = d \div r$

EXAMPLE 2 On Sunday, Ann drove 336 miles in 8 hours. What was Ann's average speed?

    STEP 1 Identify d and t.
          $d = 336$ miles $t = 8$ hours

    STEP 2 Substitute values into the rate formula: $r = d \div t$
          $336 \div 8 =$ **42 mph**

EXAMPLE 3 How long will it take Alex to complete a 105-mile bicycle race if he averages 15 miles per hour?

    STEP 1 Identify d and r.
          $d = 105$ miles $r = 15$ mph

    STEP 2 Substitute values into the time formula: $t = d \div r$
          $105 \div 15 =$ **7 hours**

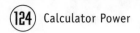
In each problem, decide whether you are trying to find the *distance*, *rate*, or *time*. Then use the correct formula to calculate your answer.

distance formula: $d = rt$	rate formula: $r = d \div t$	time formula: $t = d \div r$

1. Over a 2-day period, Dale rode a total of 14 hours on a bike ride from Portland to Seattle. If he rode a total distance of 182 miles, what average speed did he ride?

2. On the first day of her trip to Canada, June averaged 59 miles per hour for seven straight hours. What distance did June drive during this time?

3. Averaging 620 miles per hour, how long will it take an airliner to travel across the United States, a distance of 3,255 miles?

4. If she stays within the highway speed limit of 65 miles per hour, what is the farthest that Shalinda can drive in 9 hours on that highway?

5. Last Saturday, Pam was travelling for a total of 13 hours. Except for the hour she stopped for lunch and the hour for dinner, she was driving. If she drove a total of 572 miles, what was Pam's average driving speed? (**Hint:** Ask yourself, "How many hours did Pam actually drive?")

6. Brandon left home at 8:00 A.M. Tuesday. If he drove at the speed limit of 55 mph all the way, at what time did he arrive in Chicago, a distance of 275 miles from his home? (**Hint:** First find how many hours Brandon drove.)

7. What speed must Erik average if he leaves home at 1:00 P.M. and hopes to reach San Francisco by 6:00 P.M., a distance of 265 miles? (**Hint:** How many hours does Erik want to drive?)

Finding an Average

 An **average** is the total sum of a group of numbers divided by how many numbers there are. For example, suppose your gas bills during the summer months were as follows: June, $41.78; July, $46.12; and August, $38.34. Your average bill is easily computed.

To compute the average of a group of numbers, add the numbers, and then divide the sum by the number of numbers in the group.

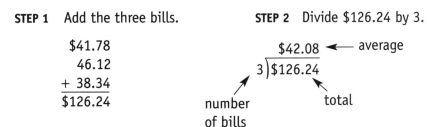

STEP 1 Add the three bills.

$41.78
46.12
+ 38.34
$126.24

STEP 2 Divide $126.24 by 3.

$42.08 ← average
3)$126.24

number of bills   total

ANSWER: $42.08

Discovery: An average is usually not equal in value to any of the numbers in the group you add. However, you'll discover that the average is often close to the middle value of the group.

 Use your calculator to answer each question below.

1. The three day-care centers near Max's home have monthly charges of $215.50, $218.00, and $212.25 for half-day child care. Determine the average cost of these centers.

2. Saturday night, Marielle bowled four games and got these scores:

   first game: 168   second game: 179
   third game: 182   fourth game: 170

   To the nearest whole number, what was Marielle's average score?

3. The price of a small pizza in Newport Beach varies quite a bit, as prices at the right show. To the nearest cent, determine the average price of a small pizza on this list.

Pizza Heaven	$14.75
Pizza Supreme	$12.80
Sparkey's Pizza	$16.00
Mamma's Pizza	$13.90
Mario's Little Italy	$14.39

Computing Simple Interest

 Interest is money that is earned (or paid) for the use of money.

- If you deposit money in a savings account, interest is money that the bank pays you for using your money.

- If you borrow money or charge purchases on a credit card, interest is money that you pay the lender or charge-card company.

Simple interest is interest on a **principal** (the original amount borrowed or deposited). To compute simple interest, use the **simple interest formula.** In words: *interest* equals *principal* times *rate* times *time*. In symbols, the formula is $i = prt$.

Interest (*i*) =	**Principal (*p*)** ×	**Rate (*r*)** ×	**Time (*t*)**
Expressed in dollars	Expressed in dollars	Expressed as a percent	Expressed in years

EXAMPLE 1   Tani deposited $600 in a savings account that pays 5.5% simple interest. How much interest will Tani's account earn in 3 years?

		Press Keys	**Display Reads**
STEP 1	Identify *p*, *r*, and *t*. $p = \$600$, $r = 5.5\%$, $t = 3$	6 0 0	600.
		×	600.
STEP 2	To find *i*, use $i = prt$. $i = 600 \times 5.5\% \times 3$	5 · 5	5.5
	When you press %, the display reads 33. So $33 is the interest earned in one year.	% *	33.
		×	33.
		3	3.
		=	99.

ANSWER: $99

Note: In actual practice, banks use a more complicated interest formula called the **compound interest formula.**

*On some calculators, you may need to press = to complete the calculation.

Although interest is earned (or paid) at a yearly rate, some deposits and loans are for part of a year. When using the simple interest formula, you change the part of a year to a decimal fraction. **Remember:** 1 month = $\frac{1}{12}$ year.

EXAMPLE 2   6 months = $\frac{1}{2}$ year

| 6 | ÷ | 1 | 2 | = | 0.5 |

ANSWER: 0.5 year

EXAMPLE 3   1 year, 8 months = $1\frac{8}{12}$ years

| 8 | ÷ | 1 | 2 | = | 0.6666666 |

ANSWER: 1.667 years

$= 0.667$

This is the part of the year rounded to the thousandths place.

Express each time below as a decimal. Round each decimal to the thousandths place.

1. 5 months

2. 3 years 1 month

3. 4 years 7 months

Solve each problem below by using the interest formula *i* = *prt*. Round each answer to the nearest cent.

4. How much interest would be earned on a deposit of $2,500 placed in a savings account for 3 years if the account pays 3% simple interest?

5. Larry borrowed $1,500 from his partner to buy stock. He agreed to repay the amount in 2 years, including 11% simple interest. At the end of 2 years, how much interest will he owe?

6. What amount of interest can Jules earn on $750 deposited for 2 years 3 months in an account that pays 4.25% simple interest?

Problems 7 and 8 refer to the chart at the right.

7. Doni deposited $375 in a new savings account at United Bank. How much will be in Doni's account at the end of 2 years 5 months. (**Hint:** Total saved = principal + interest)

8. Armand borrowed $650 from United Bank in order to buy a new TV set. Armand will repay the bank the entire amount at the end of 15 months. How much must Armand pay the bank at that time? (**Hint:** Total owed = principal + interest)

United Bank Simple Interest Accounts and Loans	
Savings Account	3.75%
Certificate of Deposit	5.82%
Car Loan	8.50%
Boat Loan	12.75%
Personal Loans	14.50%

Paying a Mortgage

 Is someone in your family making payments on a mortgage? The following example shows how the bank calculates the balance and interest each month.

EXAMPLE   The Kalachi family borrowed $78,000 on a 30-year mortgage at 7% annual interest. The bank told them that their monthly payment is $518.94. How does the bank keep track of their mortgage amount?

At the end of each month, the bank adds the interest for that month to their loan balance. (The *monthly* interest is $\frac{0.07}{12} =$ 0.0058333.) Then the bank subtracts the monthly payment of $518.94, giving a new balance for the start of the next month.

Balance to Start Month	Interest for the Month	Balance to End the Month	Monthly Payment	New Balance
78,000.00	455	78,455.00	518.94	77,936.06
77,936.06				

Then the bank calculates and adds the interest for the next month. After it receives the monthly payment, it calculates the next "new balance."

Balance to Start Month	Interest for the Month	Balance to End the Month	Monthly Payment	New Balance
78,000.00	455.00	78,455.00	518.94	77,936.06
77,936.06	454.63	78,390.69	518.94	77,871.75
77,871.75				

Calculate.

The Jamal family borrowed $89,500 on a 30-year mortgage at 7.25% annual interest. Their monthly payment is $610.55.

1. What is the monthly interest rate?

2. What is the amount of interest that they owe at the end of the first month? (**Hint:** Multiply $89,500 by your answer to problem 1.)

3. Add the interest and the amount of the loan, and then subtract the first payment. What is the balance to start the second month?

4. Lorna Sessions borrowed $48,500 on a 15-year mortgage at an annual interest of 6.75%. Her monthly payment is $429.18. Fill in the missing numbers in this chart, which tells her monthly balances for the first year of her mortgage.

Balance to Start Month	Interest for the Month	Balance to End the Month	Monthly Payment	New Balance
48,500.00		48,772.81	429.18	48,343.63
48,343.63	271.92	48,615.56		48,186.38
48,186.38		48,457.43	429.18	48,028.25
	270.16	48,298.41		47,869.23
47,869.23	269.26			47,709.31
47,709.31		47,977.67		47,548.49
47,548.49	267.46			47,386.77
		47,653.32		
47,224.14				47,060.60
47,060.60	264.72	47,325.32		
		47,159.93		46,730.75
46,730.75	262.86		429.18	46,564.43

5. Lorna's interest on her mortgage is a deduction on her income tax return. What was the total in interest that Lorna paid during the year?

6. How much of the mortgage has been paid off in 1 year? (**Hint:** Find the difference between the original balance and the final balance for the year.)

7. Look at the column that has the monthly payment. Are the amounts in this column increasing, decreasing, or staying the same?

8. Look at the column that has the interest each month. Are the amounts in this column increasing, decreasing, or staying the same?

9. Look at the column that has the balance to start each month. Are the amounts in this column increasing, decreasing, or staying the same?

Find the balance to start the second month for the following mortgages.

10. $55,000 at 6.5% annual interest; the monthly payment is $347.64.

11. $91,500 at 7.125% annual interest; the monthly payment is $616.45.

12. $138,000 at 6% annual interest; the monthly payment is $827.38.

Using Measurement Formulas

The distance around a flat object or figure is called its **perimeter** or **circumference.** The amount of a flat region covered by the object or figure is called its **area.** Here are some **measurement formulas** for circles, squares, and rectangles.

Circle

$C = 2\pi r$
$A = \pi r^2$

where C is the circumference, $\pi \approx 3.14$, r is the radius, A is the area

Square

$P = 4s$
$A = s^2$

where P is the perimeter, s is the side, A is the area

Rectangle

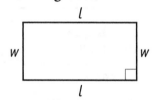

$P = 2(l + w)$
$A = lw$

where P is the perimeter, l is the length, w is the width, A is the area

Triangle

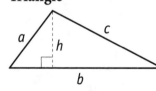

$P = a + b + c$
$A = \frac{1}{2}bh$

where P is the perimeter, a, b, and c are the sides, h is the height, A is the area

The amount of space that a solid object takes up is called its **volume.** Here are some measurement formulas for volume.

Rectangular Solid

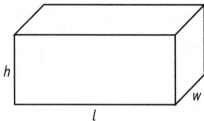

$V = l\,w\,h$
where l, w, and h are the length, width, and height

Cube

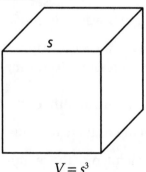

$V = s^3$
where s is the side

Cylinder

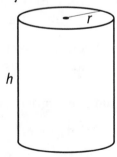

$V = \pi r^2 h$
where r is the radius and h is the height

<u>EXAMPLE 1</u> Find the circumference of a circle with an 8-inch radius. Round the answer to the nearest tenth of an inch.

In the formula $C = 2\pi r$, replace π with 3.14 and r with 8. Multiply as shown at the right.

Press Keys	Display Reads
2	2.
×	2.
3 · 1 4	3.14
×	6.28
8	8.
=	50.24

ANSWER: 50.2 inches

<u>EXAMPLE 2</u> What is the volume of a cylinder that has a radius of 1.5 feet and a height of 4.5 feet? Express your answer to the nearest tenth of a cubic foot.

In the formula $V = \pi r^2 h$, replace π by 3.14, r by 1.5, and h by 4.5.

(**Note:** Volume is measured in cubic units and area is measured in square units.)

Press Keys	Display Reads
3 · 1 4	3.14
×	3.14
1 · 5	1.5
×	4.71
1 · 5	1.5
×	7.065
4 · 5	4.5
=	31.7925

ANSWER: 31.8 cubic feet

Use the area and perimeter formulas for each figure (area and circumference for the circle). Round each answer to the nearest tenth.

1.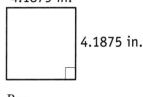

   4.1875 in.

   4.1875 in.

   $P =$
   $A =$

2.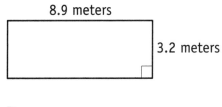

   8.9 meters

   3.2 meters

   $P =$
   $A =$

3.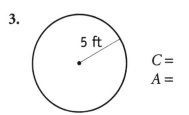

   5 ft

   $C =$
   $A =$

4.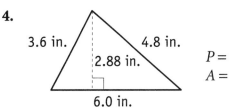

   3.6 in.   4.8 in.

   2.88 in.

   6.0 in.

   $P =$
   $A =$

Use the volume formulas for each figure. Round your answers to the nearest hundredth of a unit.

5.

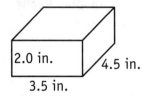

2.0 in. 4.5 in.

3.5 in.

$V =$

6.

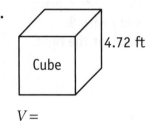

Cube 4.72 ft

$V =$

7. 4.27 yd

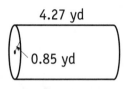

0.85 yd

$V =$

8.

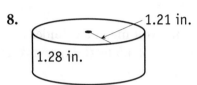

1.21 in.

1.28 in.

$V =$

For each triangle below, a square is drawn from each side of the triangle. Find the area of each of the three squares.

9.

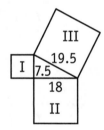

III

19.5

I 7.5

18

II

Area of square   I =
                 II =
                 III =

10.

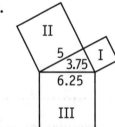

II

5

3.75 I

6.25

III

Area of square   I =
                 II =
                 III =

11. At the right, the cylinder fits tightly inside the rectangular solid. What is the difference between the volume of the rectangular solid and the volume of the cylinder? Round your answer to the nearest tenth of a cubic inch.

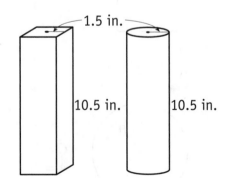

1.5 in.

10.5 in.   10.5 in.

Square Roots

The **square root** of a number is found by asking, "What number times itself equals this number?" For example, to find the square root of 36 you ask, "What number times itself equals 36?"

The answer is 6 because $6 \times 6 = 36$. The number 6 is the square root of 36.

The symbol for square root is $\sqrt{}$. For example, $\sqrt{36} = 6$.

Numbers that have whole-number square roots are called **perfect squares.** You can show the number 6 multiplied by itself as 6×6 or as 6^2 (this is read "six squared" or as "six to the second power"). A list of perfect squares is easily made by "squaring" whole numbers. The first 12 perfect squares are shown in the table below.

Table of Perfect Squares

$1^2 = 1$	$4^2 = 16$	$7^2 = 49$	$10^2 = 100$
$2^2 = 4$	$5^2 = 25$	$8^2 = 64$	$11^2 = 121$
$3^2 = 9$	$6^2 = 36$	$9^2 = 81$	$12^2 = 144$

Some four-function calculators have a square root key $\boxed{\sqrt{}}$. The square root key enables you to easily find the square root of *any* number, not just perfect squares.

<u>EXAMPLE 1</u>   What is the square root of 17?

To solve on your calculator, enter the number 17 and press the square root key.	**Press Keys**	**Display Reads**
	$\boxed{1}\ \boxed{7}$	$\boxed{17.}$
	$\boxed{\sqrt{}}$	$\boxed{4.1231056}$

ANSWER: 4.12 (rounded to the hundredths place)

Write each sentence below in symbols. The first one is an example.

1. 8 is the square root of 64. ___$8 = \sqrt{64}$___

2. 13 is the square root of 169. _____

3. 2.4 is the square root of 5.76. _____

 Use your calculator to find each square root. Round answers to the hundredths place.

4. $\sqrt{225}$ $\sqrt{400}$ $\sqrt{900}$ $\sqrt{10}$ $\sqrt{20}$

5. $\sqrt{23}$ $\sqrt{42}$ $\sqrt{15.6025}$ $\sqrt{0.746}$ $\sqrt{0.09}$

Evaluating Expressions Containing Squares and Square Roots

In some problems, squares and square roots appear together. On this page, you'll learn how you can use memory keys to find the values of such expressions.

EXAMPLE 2 Find the value of $\sqrt{11^2 + 13^2}$.

To evaluate this expression, you must first find the value of the sum of the two squares. You then find the square root of this sum.

		Press Keys	**Display Reads**
STEP 1	Compute the value 11^2. Press [M+] to place this value in memory.	[1] [1] [×] [1] [1] [*] [M+]	M 121.
STEP 2	Compute the value 13^2. Press [M+] to add this value to the number (11^2) now stored in memory. (You would use [M−] for a problem involving subtraction.)	[1] [3] [×] [1] [3] [*] [M+]	M 169..
STEP 3	Press [MR] to display the sum $11^2 + 13^2$ now stored in memory.	[MR]	M 290.
STEP 4	Press [√] to find the square root of 290.	[√]	M 17.029386

ANSWER: 17.03 (rounded to the hundredths place)

···

Calculate the value of each expression below. Round each answer to the hundredths place. Remember to clear the display and the memory before you start each problem.

6. $8^2 + 9^2 =$ $\qquad$ $4.5^2 + 6.8^2 =$ $\qquad$ $12^2 - 7^2 =$

7. $\sqrt{3^2 + 4^2}$ $\qquad$ $\sqrt{8^2 + 11^2}$ $\qquad$ $\sqrt{92^2 - 8^2}$

8. $\sqrt{14^2 + 17^2}$ $\qquad$ $\sqrt{6.7^2 + 4.2^2}$ $\qquad$ $\sqrt{92^2 - 82^2}$

*On some calculators, you may need to press [=] before [M+].

Right Triangles and the Pythagorean Theorem

On the next two pages, you'll see how memory keys can simplify problems involving the **Pythagorean theorem.** First, a short review.

A **right triangle** is a triangle in which two sides form a right angle. The side opposite the right angle is called the **hypotenuse.**

The Greek mathematician Pythagoras discovered that in a right triangle, the square of the length of the hypotenuse is equal to the sum of the squares of the lengths of the other two sides. This relationship is called the Pythagorean theorem.

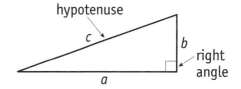

Using the labels on the triangle above, the Pythagorean theorem can be written as

$$c^2 = a^2 + b^2$$

In other words, hypotenuse squared = side squared + side squared.

EXAMPLE   What is the length of the hypotenuse (c) of the triangle pictured at the right?

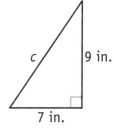

From the Pythagorean theorem, you can write

$$c^2 = 7^2 + 9^2$$

You can find c by taking the square root of the sum of the squares of the sides.

$$c = \sqrt{7^2 + 9^2}$$

To solve for c on your calculator, follow the steps shown below.

		Press Keys	Display Reads
STEP 1	Compute 7^2. Place the answer (49) in memory.	7 × 7 * M+	M 49.
STEP 2	Compute 9^2. Add this answer (81) to the memory.	9 × 9 * M+	M 81.
STEP 3	Press MR to display the sum of $7^2 + 9^2$ (49 + 81).	MR	M 130.
STEP 4	As displayed, this sum is 130. Press √ to find the square root of 130.	√	M 11.401754

ANSWER: 11.4 in. (rounded to the tenths place)

*On some calculators, you may need to press = before M+.

Using the right triangle shown here, compute the length of the hypotenuse *c*. Fill in the key symbols and the display reading for each step below.

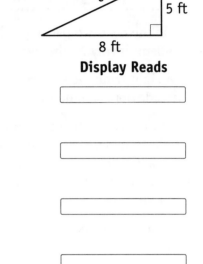

	Press Keys	**Display Reads**

1. Compute 5^2. Place the answer in memory.

2. Compute 8^2. Add this number to the memory.

3. Display the sum $5^2 + 8^2$, the sum that is now stored in memory.

4. Find the square root of the displayed number (the sum of $5^2 + 8^2$).

 Use your calculator to solve each problem below.

5. In a right triangle, one side measures 7 feet and the second side measures 10 feet. To the nearest tenth of a foot, what is the length of the hypotenuse of this triangle?

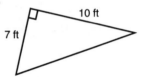

6. A ladder is leaning against the side of a house. The base of the ladder is 5 feet from the house. The top of the ladder just reaches the roof that is 13 feet above the ground. Determine the length of the ladder to the nearest tenth of a foot.

7. One side of a triangular lot is 28 yards long. A second side is 34 yards long. To the nearest yard, what is the length of the third side of this lot?

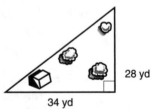

8. Alissa hiked 6.5 miles due north of her car before turning east. She then walked 3.8 miles due east before stopping to rest. At the point she rested, what was Alissa's direct distance from her car? Express your answer to the nearest tenth of a mile.

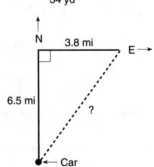

Calculator Application Review

Congratulations on completing Using Number Power. We hope you find your new skills valuable both at home and on the job.

These final three pages will give you a chance to review your calculator skills in some of the ways you're most likely to use them. Work each problem carefully with your calculator. Check your answers with the answer key.

1. Determine the daily balances in the checking account register below. Record your answers in the BALANCE column.

Record All Charges or Credits That Affect Your Account

Number	Date	Description of Transaction	Payment/ Debit (–)	✓ T	Fee (If Any) (–)	Deposit/ Credit (+)	Balance $ 641 90
309	7/6	Hillcrest Apartments	$ 275 00		$	$	
310	7/7	Value Food Store	19 85				
311	7/10	Northern Power Co.	107 33				
312		—VOID—					
313	7/14	Hair Palace	12 00				
	7/16	Deposit Paycheck				342 61	
314	7/18	2D Variety Store	27 93				

2. Complete the Total Amount column for items 2 through 5 on the purchase order form below. Then add the five entries to find the Total Purchase amount.

Wholesale Bathroom Supplies

	Item #	Description	Quantity	Cost/Per	Total Amount
1.	A 241	Bath mat	14	$18.45	$258.30
2.	B 647	Bath towel	29	$10.98	
3.	D 832	Wall mirror	7	$73.49	
4.	F 301	Shower curtain	16	$19.99	
5.	P 970	Window curtain	8	$48.75	
				Total Purchase:	

3. Which of the three stores below is offering the best buy on a case of 12 bottles of Hair Glow shampoo?

FOOD PLUS HAIR GLOW Shampoo	VALUE RITE HAIR GLOW Shampoo	DONNIE'S HAIR GLOW Shampoo
$1.89 per bottle $1.25 rebate each case	$1.79 per bottle $0.14 per bottle discount on case purchase	$1.74 per bottle "Buy 11 and get 1 free!"

4. Darrell's van can carry no more than 348 bricks at one time.

   a. How many trips will it take Darrell to move 7,500 bricks from a brickyard to a construction site?

   b. How many bricks will Darrell carry on his final trip?

5. Cecelia's gas bills for four months are shown below. Compute Cecelia's average monthly gas bill during this time. Round your answer to the nearest dollar.

   November: $158.94
   December: $183.56
   January: $192.67
   February: $174.40

6. Damian bought 6.52 pounds of tuna steaks and paid $81.50. Fill in the statement below. Calculate the cost per pound for the tuna steaks.

Number of Pounds	×	Cost per Pound	=	Total Cost
_____	×	_____	=	_____

Problems 7–10 are based on the paycheck stub below.

Smythe Electronics

Employee: Angie Lewis	Gross Pay	Federal Income Tax	State Income Tax	Social Security	Net Pay
Current Pay Period (two weeks)	$576.94	$80.77	$14.42	$40.02	$441.73
Year-to-Date	$4,038.58	$565.39	$100.94	$280.14	$3,092.11

7. How much are Angie's total deductions for the pay period shown?

8. About how much federal income tax does Angie pay each year?

9. Angie works 40 hours each week. Determine to the nearest cent her gross pay per hour.

10. What is Angie's net pay per hour (based on her 40-hour work week).

Problems 11–13 refer to the circle graph shown at the right.

11. To the nearest percent, what percent of their total take-home income do the Allisons spend on food?

12. The Allisons spend 65% of their housing expenses on rent. To the nearest dollar, how much yearly rent do the Allisons pay?

13. Next year the Allisons plan to save 15% more money than they saved this year. To do this, how much will they need to save next year?

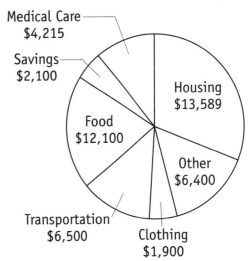

ALLISON FAMILY INCOME
Take-home Pay: $46,804

Dollar Breakdown of Total Expenses

Medical Care $4,215
Savings $2,100
Food $12,100
Housing $13,589
Other $6,400
Transportation $6,500
Clothing $1,900

Problems 14 and 15 are based on the following information.

For a family picnic, Adam bought 3.5 pounds of fruit salad at a price of $1.49 per pound and 4.5 pounds of salad greens at $1.19 per pound. Adam paid the clerk with a 20-dollar bill.

14. Which arithmetic expression shows how to calculate Adam's change?

   a. $20.00 − ($1.19 × 3.5) − ($1.49 × 4.5)
   b. $20.00 − ($1.49 × 3.5) + ($1.19 × 4.5)
   c. $20.00 − ($1.49 × 3.5) − ($1.19 × 4.5)

15. Use your calculator to find the value of the arithmetic expression you chose in problem 14.

16. As shown at the right, Edna's backyard is in the shape of a rectangle. Use the Pythagorean theorem to find the distance c. Express your answer to the nearest foot.

   (**Hint:** $c^2 = a^2 + b^2$)

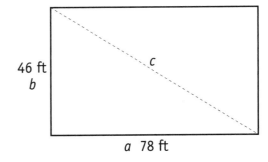

46 ft
b
c
a 78 ft

17. Fran works in a sheet metal shop that builds custom-made containers. She recently built the gasoline tank shown at the right.

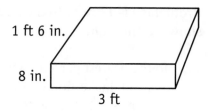

1 ft 6 in.

8 in.

3 ft

  a. Use the volume formula, $V = lwh$, to determine the container's volume to the nearest cubic foot.

  b. If 1 cubic foot holds about 7.5 gallons, how many gallons of gas can this tank hold?

Calculator Application Review Chart

Circle the number of any problem that you miss. Then return to the appropriate practice pages and review that material before redoing the problems. A passing score is 14 correct answers.

PROBLEM NUMBER	SKILL AREA	PRACTICE PAGES
1, 2, 3, 4	whole numbers and money	15–42
5, 6, 7, 8, 9, 10	decimals and fractions	45–70
11, 12, 13	percent	73–86
14, 15, 16, 17	calculator memory	89–101

Answer Key

Pages 1–3, Calculator Pretest

1. 0, 1, 2, 3, 4, 5, 6, 7, 8, 9
2. ×
3. −
4. b
5. a
6. e
7. c
8. d
9. b
10. c
11. 12
12. 8
13. 15
14. 45
15. 5
16. 2
17. 3,500 + 5,800 = 9,300
18. 440 − 80 = 360
19. 0.3
20. 0.27
21. 0.105
22. 3.002
23. 14.294
24. 14.3
25. $4\frac{3}{5}$
26. 14.324
27. 25
28. 60%
29. 100
30. 35
31. 7
32. 23
33. 130
34. 5
35. 544

Pages 9–10

Answers will vary. For calculators similar to the example on page 7, the answers are as follows:

1. Solar powered
2. The ON/OFF key
3. 0.
4. eight digits, 1 through 8
5. 99,999,999
6. Press C once.
7. 2.57; 17.81; 20.56; 0.49; 0.08; 0.10
8. 4.19; 6.12
9. 0.27; 0.42
10. 0.01; 0.08
11. 1.06; 2.07

Page 12

1. 2,450
2. 875 (no comma)
3. 4,056
4. 39,450
5. 1,832
6. 29,609
7. $18.32 (no comma)
8. $2,471.60
9. $649.09 (no comma)
10. $5,038.18
11. **b.** three hundred forty
12. **c.** nine hundred
13. **c.** three dollars and seven cents
14. **a.** twenty cents
15. 95.
16. 243.
17. 3529.
18. 8406.
19. 15.82
20. 204.09

Page 13

1. O, I, Z, E, h, S, g, L, B
2. O, I, E
3. *hi, she, oil*
4. Some examples are *he, go, is,* and *hi.*
5. Some examples are *oil, she, his, bog,* and *log.*

Page 16

1. 8 + 4 =
2. 3 9 + 1 7 =
3. 2 0 6 3 + 9 8 9 =
4. 5 · 0 9 + 4 · 2 6 =
5. 1 2 8 + 8 9 =
6. 7 · 9 0 + 3 · 7 8 =
7. 3 5 + 1 4 =
8. 1 5 7 + 6 1 =
9. 5 2 0 9 + 2 4 0 0 =
10. 4 · 5 0 + 6 · 1 2 =
11. 59; 657; 2,822
12. $1.65; $55.07; $415.74
13. 5,418 pounds

Page 17

1. [2] [7] [+] [1] [8] [=]
2. [2] [0] [+] [3] [9] [=]
3. [1] [5] [3] [+] [8] [9] [=]
4. [.] [6] [7] [+] [.] [4] [9] [=]

	Wrong Key	Double Keying	Transposed Digits
1.		✓	
2.			✓
3.	✓		
4.			✓

Page 18

1. 6	7. 8	13. 99	19. 757
2. 13	8. 17	14. 106	20. 1,106
3. 9	9. 50	15. 88	21. 800
4. 11	10. 91	16. 133	22. $1,413
5. 7	11. 97	17. 469	23. $6.68
6. 13	12. 74	18. 1,127	24. $11.51

25–26. Answers will vary on categorizing answers. Many students, though, will choose to do the odd-numbered problems in their heads or with pencil and paper. These problems involve no carrying. Many students prefer to do carrying problems with a calculator, especially when more than single-digit problems are involved.

Page 20

1. [1] [2] [+] [9] [+] [8] [=]
2. [3] [5] [+] [2] [7] [+] [9] [=]
3. [1] [.] [5] [3] [+] [.] [9] [4]
   [+] [.] [5] [8] [=]

Problems 6, 7, 8, and 11 are incorrect.

6. $12.98   8. 390

7. $635   11. 8,001

Page 22

1. [8] [6] [−] [4] [9] [=]
2. [5] [0] [8] [−] [2] [1] [7] [=]
3. [1] [0] [.] [0] [0] [−] [7] [.] [9] [2] [=]
4. [8] [.] [0] [0] [−] [3] [.] [1] [1] [=]
5. Example:
   [1] [0] [−] [7] [.] [9] [2] [=] ;
   [8] [−] [3] [.] [1] [1] [=]

Incorrect problems are 6, 8, and 10.

6. 18	12. $2.92; $11.40
8. 316	13. 1,269 pounds
10. $41.54	14. $3.95
11. 15; 34; 91	

Pages 23–24

1. mental math	5. mental math
2. mental math	6. A
3. mental math	7. larger numbers
4. calculator	8. Answers will vary.

Page 26

1.
$$\begin{array}{r} 90 \\ +\ 70 \\ \hline 160 \end{array} \qquad \begin{array}{r} 50 \\ +30 \\ \hline 80 \end{array} \qquad \begin{array}{r} 90 \\ -40 \\ \hline 50 \end{array}$$

2.
$$\begin{array}{r} 500 \\ -200 \\ \hline 300 \end{array} \qquad \begin{array}{r} 900 \\ +\ 100 \\ \hline 1,000 \end{array} \qquad \begin{array}{r} \$700 \\ -\ 300 \\ \hline \$400 \end{array}$$

3.
$$\begin{array}{r} 5,000 \\ +2,000 \\ \hline 7,000 \end{array} \qquad \begin{array}{r} 8,000 \\ +\ 5,000 \\ \hline 13,000 \end{array} \qquad \begin{array}{r} 12,000 \\ -\ 8,000 \\ \hline 4,000 \end{array}$$

4. d. 90   6. a. 50   8. c. 700

5. e. 60   7. b. 100

Pages 27–28

1. b. $3.81

2. c. 8,019

3. a. $18.74

Estimates will vary. Example answers are given below.

	Exact	Estimate
4.	892 calories	900 calories
5.	696 miles	700 miles
6.	118 miles	100 miles
7.	$56.11	$56.00
8.	42	40
9.	$177.60	$180.00

Pages 29–30

1 and 2 (See completed register below.)

Number	Date	Description of Transaction	Payment/ Debit (–)		✓ T	Fee (–)	Deposit/ Credit (+)		Balance $1568 43	
202	6/1	North Street Apartments	$ 825	00	✓	$	$		743	43
203	6/4	Amy's Market	39	74					703	69
204	6/8	Import Auto Repair	109	66	✓				594	03
	6/15	Payroll Deposit					1442	45	2036	48
205	6/19	Value Pharmacy	13	29	✓				2023	19
206	6/21	Gazette Times	18	75					2004	44
207	6/23	Nelsen's	39	83	✓				1964	61
208	6/24	Hi-Ho Foods	63	79	✓				1900	82
	6/26	Check Reorder Charges	26	50					1874	32
	6/30	Payroll Deposit					1442	45	3316	77

3. **STEP 1** (See completed register above.)

   STEP 2 Check #203 $39.74
             Check #206   + 18.75
                               $58.49

   STEP 3 06/30 Statement Balance $3369.76
                                        –   58.49
                                          $3311.27

   STEP 4 06/30 Register Balance $3316.77
             Bank Service Charge   –   5.50
                                          $3311.27

   STEP 5 The amounts computed in Steps 3 and 4 are both equal to $3311.27.

Page 32

1. ⎡7⎤ ⎡6⎤ ⎡×⎤ ⎡4⎤ ⎡0⎤ ⎡=⎤
2. ⎡1⎤ ⎡0⎤ ⎡6⎤ ⎡×⎤ ⎡8⎤ ⎡8⎤ ⎡=⎤
3. ⎡9⎤ ⎡.⎤ ⎡0⎤ ⎡4⎤ ⎡×⎤ ⎡7⎤ ⎡=⎤

4. **Exact**   **Estimate**   5. **Exact**   **Estimate**
   1,536   1,500   10,944   12,000
   3,953   4,200   9,248   9,000
   1,672   1,800   15,318   14,000

6. 280; 912; 1,955   8. $140.70; $254.80; $937.62

7. 6,216; 2,002; 98,298   9. $85.44

Page 34

1. dividend = 152; divisor = 19

   ⎡1⎤ ⎡5⎤ ⎡2⎤ ⎡÷⎤ ⎡1⎤ ⎡9⎤ ⎡=⎤

2. dividend = $19.68; divisor = 8

   ⎡1⎤ ⎡9⎤ ⎡.⎤ ⎡6⎤ ⎡8⎤ ⎡÷⎤ ⎡8⎤ ⎡=⎤

3. dividend = 361.56; divisor = 23

   ⎡3⎤ ⎡6⎤ ⎡1⎤ ⎡.⎤ ⎡5⎤ ⎡6⎤ ⎡÷⎤ ⎡2⎤ ⎡3⎤ ⎡=⎤

4. dividend = 576; divisor = 27

   ⎡5⎤ ⎡7⎤ ⎡6⎤ ⎡÷⎤ ⎡2⎤ ⎡7⎤ ⎡=⎤

5. Exact: $418 \div 19 = 22$; $693 \div 21 = 33$; $288 \div 18 = 16$

   Estimate: $400 \div 20 = 20$; $700 \div 20 = 35$; $300 \div 20 = 15$

6. 45; 26; $48; $61

7. 32 tables

8. $35

Pages 35–36

1. **b.** division: $15   3. **a.** multiplication: $19.60

2. **a.** multiplication: 128   4. **b.** division: 32 min

   Estimates may vary. Example estimates are given.

5. Estimate: $\$9.00 \times 40 = \$360.00$
   Exact: $\$8.75 \times 39 = \$341.25$

6. Estimate: $\$20,000,000 \div 10 = \$2,000,000$
   Exact: $\$19,972,800 \div 9 = \$2,219,200$

7. Estimate: $4,000 \div 20 = 200$
   Exact: $4,392 \div 18 = 244$

8. Estimate: $7,000 \times \$5.00 = \$35,000$
   Exact: $7,090 \times \$4.75 = \$33,677.50$

9. $268 \times 52 = \$13,936$ **11.** $\$330 \div 40 = \$8.25$

10. $\$330 \times 52 = \$17,160$ **12.** $\$34,580 \div 52 = \$665$

Pages 37–38

1. cost of hats $= 3 \times \$14.89 = \$44.67$
   change $= \$50.00 - \$44.67 = \$5.33$

2. eggs in each case $= 24 \times 12 = 288$
   cases used $= 1,728 \div 288 = 6$

3. $23.35 (\$15.75 + \$0.95 \times 8)$

4. $121.75 (\$28.75 \times 3 + \$17.75 \times 2)$

5. 11 (subtract 119 from 405 and divide by 26)

6. $1,363.09 (subtract the Total, \$83,920.19, from \$85,283.28)

Page 39

CASH	CURRENCY	2803	
	COIN	70	50
LIST CHECKS SINGLY			
24 - 12		44	89
31 - 29		39	85
42 - 16		120	00
TOTAL FROM OTHER SIDE		—	
TOTAL		3078	24
LESS CASH RECEIVED		0	
NET DEPOSIT		3078	24

Page 40

1. $4\ 7\ 5 \div 8\ 0 = 5.9375$
   5 full buses, 1 more bus = 6 buses

2. $4\ 5\ 0 \div 3\ 5 = 12.857143$
   12 full disks, 1 more disk = 13 disks

Page 42

1. **STEP 1** $598 \div 27 = 22.148148$
   STEP 2 $22 \times 27 = 594$
   STEP 3 $598 - 594 = 4$
   Whole number plus remainder $= 22$ r 4

2. 15 r3

3. 9 r17

4. 18 r11

5. 39 r5

6. $114 \div 12 = 9.5$
   The last load will contain 6 cubic yards.

7. $3,185 \div 350 = 9.1$
   She drove 35 miles on the final day.

Page 43

7	0	3	■	2	0
1	■	9	3	1	
4	0	■	5	■	
■		8	0	0	
3	6	3	■	1	0
0	■	0		9	

Page 46

1. five hundredths
2. five tenths
3. five millionths
4. five thousandths
5. five hundred-thousandths
6. five ten-thousandths
7. c. 206 thousandths
8. c. 47 thousandths
9. a. 7 tenths
10. a. 15 hundredths
11. b. 9 thousandths
12. b. 8 hundredths
13. $1 \div 8 = 0.125$
   b. 125 thousandths
14. $9 \div 100 = 0.09$
   b. 9 hundredths
15. $30 \div 8 = 3.75$
   b. 3 and 75 hundredths
16. $5 \div 16 = 0.3125$
   c. 3,125 ten-thousandths
17. $20 \div 80 = 0.25$
   c. 25 hundredths
18. $245 \div 56 = 4.375$
   a. 4 and 375 thousandths

Page 48

1. $0.30
2. $1.00
3. $7.90
4. $0.60
5. $3.60
6. $4.10
7. 3.5; 3.46
8. 8.01; 8.007
9. 14.4; 14.38
10. 24.00; 24.005
11. 38.9; 38.95
12. 90.83; 90.831
13. 40.1; 40.05
14. 72.38; 72.385

15. $1.2307692 \approx 1.2; 1.375 \approx 1.4; 3.4285714 \approx 3.4$

16. $\$1.9375 \approx \$1.94; 1.875 \approx 1.88; \$1.8888888 \approx \$1.89$

17. $1.4375 \approx 1.438; 2.0454545 \approx 2.045; 2.2857143 \approx 2.286$

18. $33 \div 7 = 4.7142857 \approx 4.71$ inches

19. $2.54 \times 36 = 91.44 \approx 91.4$ centimeters

Page 50

1. 1.25 terminating

2. 0.333 . . . repeating

3. 0.1515 . . . repeating

4. 0.875 terminating

5. 1.5454545
   repeating digits: .54 . . .

6. 1.3333333
   repeating digits: .3 . . .

7. 2.5555555
   repeating digits: .5 . . .

8. 0.2424242
   repeating digits: .24 . . .

9. 3 ÷ 7 = 0.4285714.
   Carried to 12 decimal places, 3 ÷ 7 = 0.428571428571.
   5 ÷ 17 = 0.2941176. Carried to 32 decimal places,
   5 ÷ 17 = 0.29411764705882352941176470588235.

Page 51

1. ■

2. ⊖

3. ⬡

4. ☐

5. eight

6. third

7. smallest

8. best

9. 21

10. X

11. 13

12. U

13. 46,046
    52,052
    28,028
    *34,034
    *16,016

14. 1.1111111
    1.2222222
    1.3333333
    *1.4444444
    *1.6666666

15. 222
    333
    444
    *555
    *777

16. 0.3636363
    0.3939393
    0.4242424
    *0.4545454
    *0.5151515

Page 52

1. ÷ $510 ÷ 12 = $42.50

2. + $148,795 + $17,980 = $166,775

3. − 236 − 199 = 37 pounds

4. × 22 × 16 = 352 miles

5. ÷ 142 ÷ 5 = 28.4 pounds

Page 54

1. 1.36; 0.99; 3.95

2. 0.16; 2.5; 11.65

3.

Exact	Estimate
8.3	8
+ 5.6	+ 6
13.9	14

Exact	Estimate
9.7	10
+ 6.5	+ 7
16.2	17

Exact	Estimate
6.3	6
+ 7.1	+ 7
13.4	13

4.

Exact	Estimate
25.8	30
+ 13.6	+ 10
39.4	40

Exact	Estimate
75.60	80
+ 24.83	+ 20
100.43	100

Exact	Estimate
27.85	30
+ 12.4	+ 10
40.25	40

5.

Exact	Estimate
328.65	300
211.06	200
+ 123	+100
662.71	600

Exact	Estimate
547.69	500
243.63	200
+ 181.8	+200
973.12	900

Exact	Estimate
900	900
375.8	400
+ 245.18	+ 200
1,520.98	1,500

6. 37; 95; $204

7. 387; 318; $58

8. 3.74 inches

Page 56

1. 0.81

2. 0.11

3. 0.035

4. 0.23; 0.175; 0.019

5. 0.044; 0.01; 0.026

6. 0.153

7. $4.65

8. $0.70

9. 5.4°F

Pages 57–58

1. **b.** 4.83 (3 + 2 = 5)

2. **a.** 1.41 (6 − 5 = 1)

3. **c.** $21.90 ($60 − $40 = $20)

Estimates may vary.

4. Exact: 14.26 − 11.89 = 2.37 seconds
   Estimate: 14 − 12 = 2 seconds

5. Exact: 30.2 − 27.8 = 2.4 miles per gallon
   Estimate: 30 − 28 = 2 miles per gallon

6. Exact: $11.88 + $2.29 = $14.17
   Estimate: $12.00 + $2.00 = $14.00

7. Exact: 4.18 − 3.29 = 0.89 mile
   Estimate: 4 − 3 = 1 mile

8. Exact: 20.7 − 18.9 = 1.8 gallons
   Estimate: 21 − 19 = 2 gallons

9. Exact: 1.375 − 0.9375 = 0.4375
   Estimate: 1.4 − 0.9 = 0.5

Page 60

1. 46.8; 7.71; 5.628; 72.05; 0.7688

2.

Exact	Estimate
6.8	7
× 4.2	× 4
28.56	28

3.

Exact	Estimate
10.3	10
× 5.9	× 6
60.77	60

Exact	Estimate		Exact	Estimate
8.12	8		12.9	13
× .94	× 1		× 1.09	× 1
7.6328	8		14.061	13

Exact	Estimate		Exact	Estimate
4.03	4		15.4	15
× 2.1	× 2		× 2.19	× 2
8.463	8		33.726	30

4. 144; 454.2

5. 345; 554.4

6. $111.32; $283.89

7. $20.32

Page 62

1. dividend 6.54; divisor 3.1

2. dividend 8; divisor 2.7

3. b. 8 (48 ÷ 6 = 8)

4. a. 8 (32 ÷ 4 = 8)

5. b. 10 (100 ÷ 10 = 10)

6. c. 7 (56 ÷ 8 = 7)

7. 44.8; 24.3; 50.4; 9.5

8. 33.91; $3.09; $23.81; 2.41

9. $5.65

Pages 63–64

1. c. between $20 and $22

2. b. between 250 and 350

3. b. between $77 and $96

4. c. between 26 and 32

5. a. between $3 and $4

6. Exact: $3.13 ÷ 2.9 = $1.08 (to the nearest cent)
Estimate: $3.00 ÷ 3 = $1.00

7. Exact: 59.8 ÷ 22 = 2.7 pounds (to the nearest tenth of a pound)
Estimate: 60 ÷ 20 = 3 pounds

8. Exact: 0.9375 × 8 = 7.5 inches
Estimate: 1 × 8 = 8 inches

9. Exact: 48 × 8.3 = 398.4 pounds
Estimate: 50 × 8 = 400 pounds

10. $9.44 × 40 = $377.60

11. $9.44 × 1.5 = $14.16

12. $377.60 + (5 × $14.16) = $448.40

13. $425.50 ÷ 40 = $10.64 (to the nearest cent)

Page 65

At Super Kids	At Huggy Bear
1. $425.50	$472
2. $781	$814
3. $1,206.50	$1,286

4. Super Kids would be $79.50 less for three months.

Page 67

1. 0.76; $\frac{5}{8}$; 0.4; 0.865

2. $\frac{2}{3}$; $\frac{9}{17}$; $\frac{2}{3}$; $\frac{8}{10}$

3. $\frac{3}{8}$ of a dollar

4. Bit C: $\frac{13}{64}$ inch

5. No

Page 69

1. c. $8\frac{5}{24}$

2. b. $14\frac{2}{3}$

3. a. $6\frac{2}{35}$

4. b. $2\frac{4}{15}$

5. $1.39 per pound

6. $11.06

Page 70

1. $36\frac{1}{7}, 7\frac{2}{19}, 11\frac{14}{27}$

2. $24\frac{6}{17}, 29\frac{9}{23}, 45\frac{5}{11}$

3. $25\frac{1}{16}, 4\frac{43}{95}, 6\frac{33}{112}$

4. $43\frac{6}{23}, 11\frac{47}{73}, 60\frac{6}{11}$

Page 71

Letters indicate which problem has an answer that matches the answer appearing within each square.

h	n	k
l	b	c
j	f	p

Page 74

1. percent, division

2. part, multiplication

3. whole, division

Page 76

1. 18, 84, $2.04, 66

2. 368, $41.40, 274, $1.42

3. $0.02, 0.595, $1.65, 2.108

4. 45, $174, 135, $112

5. $36.54
[2] [4] [3] [.] [6] [0] [×] [1] [5] [%]

6. $11.50 [4] [6] [0] [×] [2] [.] [5] [%]

7. $55.93 [1] [5] [9] [8] [×] [3] [.] [5] [%]

Page 78

1. 20% **4.** 40%

2. 80% **5.** 32%

3. 38% **6.** 51.7%

7. 25%; 60%; 45%; 70%; 40%

8. 37.5%; 31.3%; 33.3%; 9.4%; 66.7%

9. 16% [4] [8] [÷] [3] [0] [0] [%]

10. 40% [2] [÷] [5] [%]

*Page 80

1. 268 **5.** 250

2. 1,000 **6.** $2,000

3. $28 **7.** $40

4. 150 **8.** $2,500

9. 20 pounds [6] [÷] [3] [0] [%]

10. $660 [9] [.] [9] [0] [÷] [1] [.] [5] [%]

11. $18 [1] [2] [.] [6] [0] [÷] [7] [0] [%]

*Page 82

1. 99¢ per pound [9] [0] [+] [1] [0] [%]

2. 60,264 [4] [8] [6] [0] [0] [+] [2] [4] [%]

3. $12,290.28
[1] [3] [4] [3] [2] [–] [8] [.] [5] [%]

4. $129.22
[1] [8] [4] [.] [6] [0] [–] [3] [0] [%]

5. $62.79 [5] [4] [.] [6] [0] [+] [1] [5] [%]

6. $72.80 [5] [6] [+] [3] [0] [%]

7. $112.35
[1] [4] [9] [.] [8] [0] [–] [2] [5] [%]

8. $23.54 [2] [6] [.] [1] [6] [–] [1] [0] [%]

Pages 83–84

1. b. exactly 42 (100% means all.)

2. a. less than 100% (18 is a part of 24.)

3. c. more than $860 (Shelley is getting a raise.)

4. a. less than 100% (14 is a part of 34.)

5. whole; $320 [4] [8] [÷] [1] [5] [%]

6. percent decrease; 60 [7] [5] [–] [2] [0] [%]

7. percent; 20%
[4] [2] [6] [÷] [2] [1] [5] [4] [%]

8. percent decrease; $67.20
[8] [4] [–] [2] [0] [%]

9. whole; $36.00
[2] [8] [.] [8] [0] [÷] [8] [0] [%]

10. percent; 10%
[1] [.] [2] [5] [÷] [1] [2] [.] [5] [0] [%]
since $12.50 – $11.25 = $1.25

11. percent; 13% [2] [0] [÷] [1] [5] [0] [%]
since 150 – 130 = 20

12. percent decrease; 144 pounds
[1] [5] [0] [–] [4] [%]

Page 86

1. The percents are as follows:
Housing 32%
Other 17%
Clothing 7%
Transportation 12%
Food 18%
Savings 5%
Medical Care 9%

2. $7,139.84

3. 44%

4. 23%

5. 22%

6. $558

7. $803.25

8. $2,788.75

Page 87

1. d	**Actual**	**Actual**
2. a	**6.** 100	**11.** 20
3. c	**7.** 50	**12.** 200
4. e	**8.** 5	**13.** 150
5. b	**9.** 100	**14.** 15
	10. 200	

Page 90

1. e **2. d** **3. a** **4. b** **5. c**

Page 92

1. Add 26 and 14; then subtract 9.

2. Multiply 13 × 4; then subtract 18.

3. Divide 64 by 8; then add 11.

4. Multiply 9 × 8; then add 21.

5. Divide 56 by 7; then subtract this quotient from 60.

6. Add the product of 7 × 6 to the product of 8 × 3

7. Subtract the product of 7 × 6 from the product of 9 × 8.

8. Add 7 and 6; then multiply the sum by 9.

9. Subtract 10 from 12; then multiply the difference by 8.

10. Subtract 12 from 30; then divide the difference by 6.

*On some calculators, you may need to press [=] after [%] to complete the calculation.

Pages 93–94

1. 675
2. 4
3. 21.125
4. 612

*Pages 95–96

1. 25 [2][8][−][1][7][+][1][4][=]
2. $17.73 [1][2][·][4][3][+][9][·][3][6][−][4][·][0][6][=]
3. 181 [3][9][6][−][2][0][9][−][4][3][+][3][7][=]
4. 184 [2][0][7][+][1][1][1][−][7][3][−][6][1][=]
5. $8.25 [1][3][·][4][6][−][2][·][0][9][−][3][·][1][2][=]
6. $1.12 [4][·][2][5][−][2][·][1][9][−][1][·][1][8][+][·][2][4][=]
7. 600 [4][3][+][7][×][1][2][=]
8. 7 [5][6][−][2][8][÷][4][=]
9. $20.61 [4][·][5][6][+][2][·][3][1][×][3][=]
10. 6 [4][5][+][2][8][−][1][9][÷][9][=]
11. 392 [6][3][−][3][5][×][1][4][=]
12. 80 [2][9][−][1][3][×][5][=]
13. $4.55 [6][·][2][5][+][2][·][8][5][÷][2][=]
14. 114 [1][1][2][+][1][0][9][+][1][2][1][÷][3][=]
15. 210 [1][5][6][M+][1][8][×][3][M+][MR]
16. 17 [9][M+][1][0][4][÷][1][3][M+][MR]
17. 31.3 [1][8][·][7][M+][4][·][2][×][3][M+][MR]
18. 67 [7][·][2][×][6][M+][3][·][4][×][7][M+][MR]
19. 138 [8][4][×][3][M+][5][7][×][2][M−][MR]
20. 25 [3][7][M+][3][×][4][M−][MR]
21. 6 [1][5][M+][1][5][3][÷][1][7][M−][MR]
22. 12 [2][7][·][6][M+][7][·][8][×][2][M−][MR]

23. $29.51 [3][·][7][9][×][4][M+][2][·][8][7][×][5][M+][MR]
24. $5.38 [5][0][M+][1][3][·][4][9][×][2][M−][5][·][8][8][×][3][M−][MR]
25. $6.88 [7][·][5][6][+][2][·][8][9][−][3][·][5][7][=]
26. $9.35 [8][·][2][9][−][6][·][4][2][×][5][=]
27. $17.83 [2][·][4][9][×][3][M+][5][·][1][8][×][2][M+][MR]
28. $5.12 [1][0][−][2][·][9][9][−][1][·][8][9][=]
29. 1,122 [1][4][3][+][2][3][1][×][3][=]
30. 65.1 [5][·][7][+][3][·][6][×][7][=]
31. $5.07 [2][7][·][4][6][−][1][7][·][3][2][÷][2][=]
32. $14.44 [1][4][·][6][8][+][1][6][·][2][0][+][1][2][·][4][4][÷][3][=]
33. $11.60 [5][·][8][9][−][2][·][9][9][×][4][=]
34. $19.03 [2][5][M+][5][·][2][5][×][3][M−][4][·][8][9][×][2][M+][MR]

Page 97–99

1. **b.** $25.00 − $14.79 − $4.99 = $5.22
2. **c.** ($13.95 + $1.99 + $1.99) ÷ 3 = $5.98
3. **c.** $152.90 − $41.49 + $85.00 = $196.41
4. **b.** ($337.40 − $50.00) ÷ 12 = $23.95
5. **a.** (9 + 8) × 11.5 = 195.5
6. **b.** (3 × 5) + (2 × 3) = 21
7. $84.29 − $18.75 + $125.94 = $191.48
8. $241.85 + $51.38 − $53.00 = $240.23
9. (1.9 + 1.9) × 5 = 19 miles
10. 20 − (4.3 × 4) = 2.8 pounds
11. ($329.99 − $44.99) ÷ 3 = $95.00
12. ($8.75 × 6.4) − ($3.08 × 4.5) = $42.14
13. (182 + 175 + 193) ÷ 3 = 183

*On some calculators, you may have to press [=] before you press [M+] or [M−].

*Pages 100–101

1. 8% `9` `.` `3` `0` `−` `8` `.` `6` `0` `÷` `8` `.` `6` `0` `%`

2. 27% `3` `8` `.` `5` `0` `−` `2` `7` `.` `9` `5` `÷` `3` `8` `.` `5` `0` `%`

3. 29% `6` `2` `8` `0` `0` `−` `4` `8` `6` `0` `0` `÷` `4` `8` `6` `0` `0` `%`

4. 15% `7` `2` `6` `0` `0` `−` `6` `1` `7` `1` `0` `÷` `7` `2` `6` `0` `0` `%`

5. 25% `1` `.` `5` `−` `1` `.` `2` `÷` `1` `.` `2` `%`

6. 22% `2` `.` `2` `5` `−` `1` `.` `7` `5` `÷` `2` `.` `2` `5` `%`

7. 30.1% `4` `5` `.` `8` `1` `3` `−` `3` `5` `.` `2` `1` `3` `M+` `=` `÷` `MR` `=`

8. 1.5% `8` `8` `.` `2` `7` `−` `8` `6` `.` `9` `5` `M+` `=` `÷` `MR` `=`

9. 11.6% `1` `9` `.` `5` `8` `8` `−` `2` `2` `.` `1` `6` `6` `M+` `=` `÷` `MR` `=`

10. 16.6% `1` `6` `5` `.` `2` `9` `8` `−` `1` `4` `1` `.` `7` `2` `5` `M+` `=` `÷` `MR` `=`

11. 12.4% `3` `4` `.` `2` `1` `5` `−` `3` `9` `.` `0` `7` `M+` `=` `÷` `MR` `=`

12. 57.3% `1` `3` `5` `.` `2` `8` `−` `8` `6` `M+` `=` `÷` `MR` `=`

13. 32.2% `3` `6` `.` `4` `3` `2` `−` `2` `7` `.` `5` `5` `M+` `=` `÷` `MR` `=`

14. 7.0% `9` `1` `.` `7` `5` `3` `−` `9` `8` `.` `6` `5` `2` `M+` `=` `÷` `MR` `=`

15. 30.2% `5` `7` `.` `4` `7` `7` `−` `4` `4` `.` `1` `4` `M+` `=` `÷` `MR` `=`

16. 3.7% `1` `0` `5` `.` `9` `7` `2` `−` `1` `0` `2` `.` `1` `5` `M+` `=` `÷` `MR` `=`

Pages 102–105, Posttest A

1. **c.** add two numbers
2. **d.** divide two numbers
3. **b.** add a number to memory
4. **a.** recall a stored number
5. `          3570.`
6. `          17.08`
7. `2` `4` `0` `0` `+` `7` `5` `0` `=`
8. `2` `0` `0` `−` `1` `2` `5` `−` `4` `6` `=`
9. `4` `2` `3` `.` `7` `5` `×` `6` `=`
10. `3` `6` `.` `5` `0` `÷` `5` `=`
11. `7` `÷` `8` `=`

For answers 12–15, some calculators require that you press `=` at the end.

12. `9` `0` `0` `×` `3` `5` `%`
13. `1` `5` `÷` `8` `0` `%`
14. `2` `4` `÷` `3` `0` `%`
15. `1` `6` `×` `5` `M+` `1` `4` `×` `3` `M−` `MR`
16. $214 - 26 =$ **188 pounds**
17. $35 + 17 + 24 =$ **76 pieces of clothing**
18. $\$30 - \$26.25 =$ **\$3.75**
19. $\$0.59 \times 24 \times 3 =$ **\$42.48**
20. $\$448.75 \div 38 =$ **\$11.81**
21. $170 \div 8 =$ **22 plates** (rounded up)
22. $2.5 \div 7 =$ **0.357 inches**
23. $\$1,234 \times 20\% =$ **\$246.80**
24. $\$205 \div \$1,234 =$ **17%**
25. $\$124 - 35\%$ of $124 =$ **\$80.60**
26. $210 \div 15\% =$ **\$1,400**
27. $\$50 - (\$8.98 \times 4) =$ **\$14.08**
28. $(\$2.89 \times 3) + (\$1.79 \times 5) =$ **\$17.62**
29. $(\$8.25 - \$7.50) \div \$7.50 =$ **10%**
30. $(\$350 - \$250) \div \$350 =$ **29%**

Pages 107–111, Posttest B

1. **b.** `−`
2. **c.** `M+`
3. **b.** division
4. **d.** recalling a number stored in memory

*On some calculators, you may need to press `=` to complete the calculation.

5. a. comma key

6. e. | 4 6 2 3 .

7. d. | 3 2 . 5

8. b. | 0 . 0 7 8

9. d. [3] [5] [0] [+] [1] [3] [9] [=]

10. a. [4] [6] [·] [5] [3] [÷] [3] [=]

11. e. [3] [÷] [4] [=]

12. c. [7] [6] [×] [1] [8] [%]

13. b. [3] [2] [÷] [2] [5] [6] [%]

14. e. [1] [8] [÷] [1] [6] [%]

15. c. [1] [4] [×] [7] [M+] [2] [5] [×] [6] [M+] [MR]

16. a. $10 - $3.95 = **$6.05**

17. d. $14.95 + $3.45 + $3.75 = **$22.15**

18. a. $20 - 3($5.79) = **$2.63**

19. c. $24 \times 12 \times 3 =$ **864**

20. b. $29.40 \div 19 =$ **$1.55**

21. d. $28 \div 5 =$ **6** (rounded up)

22. d. $43 \div 6 =$ **7.167** (rounded to the nearest thousandth)

23. a. $26,850 \times 38\% =$ **10,203**

24. b. $19 \div 178 =$ **11%** (rounded to the nearest percent)

25. c. $425 + (8\% \times 425) =$ **459**

26. e. $720 \div 29\% =$ **2,500** (rounded to the nearest hundred)

27. a. $394 - (58 \times 3) =$ **220**

28. d. ($12.48 \times 6) + ($4.29 \times 30) =$ **$87.75**

29. b. ($137,500 - $125,000) \div $125,000 =$ **10%**

30. d. ($14.75 - $12.50) \div $14.74 =$ **15%** (rounded to the nearest percent)

Page 114

1. Daily Mileage

Mon.	382
Tue.	407
Wed.	354
Thu.	430
Fri.	288
Sat.	376
Sun.	529
TOTAL	2,766

2. WEEKLY TOTAL: 2,766 miles

3. a. Sat.–Sun. = 905 miles

   b. Mon.–Fri. = 1,861 miles

Page 115

1. Drills: $389.40
Chests: $199.95
Hammers: $116.55
Shovels: $320.96
Vises: $123.92
Ladders: $337.74
Screwdrivers: $154.40

2. TOTAL PURCHASE: $1,642.92

3. $1,526.37 ($1,642.92 − $116.55)

Page 116

1. a. Subtract $539.57 from $684.80.
   b. Add $79.24, $14.27, and $51.72.

2. $8,217.60

3. $79.24

4. $145.23 ($684.80 − $539.57)

5. 12 pay periods ($8,217.60 ÷ $684.80)

6. a. $16.00 (to the nearest cent; $684.80 ÷ 42.8)
   b. $12.61 (to the nearest cent; $539.57 ÷ 42.8)

7. $1,344.72 ($51.72 × 26)

Page 117

	Chevrolet	Ford
1.	$900.00	$992.00
2.	$10,466.31	$9,638.93
3.	706	857
4.	$1,037.82	1,259.79
5.	$74.85	$74.85
6.	$1,818.67	$2,191.64
7.	$12,284.98	$11,830.57

*Pages 118–119

Name	M	T	W	T	F	S	S	①Total Hours	②Regular Hours	③Overtime Hours
Allen	8.0	8.0	9.0	9.5	7.5	3.0		45	40	5
Cook	8.0	9.5	6.5	9.0	8.5	2.5	2.5	46.5	40	6.5
Dart	7.5	9.0	5.5	7.5	6.5			36	36	0
Franks		8.5	7.5	8.5	9.0	8.5	3.5	45.5	40	5.5
Norris	9.0	8.5	8.0	5.5	7.5	8.5		47	40	7

*On some calculators, you may have to press [=] to complete the calculation.

Name	① Regular Hours	② Regular Pay Rate	③ Total Regular Pay	④ Overtime Hours	⑤ Overtime Pay Rate	⑥ Total Overtime Pay	Total Pay	⑧ Total % Withholding	⑨ Net Pay
Allen	40	$8.32	$332.80	5	$12.48	$62.40	$395.20	18%	$324.06
Cook	40	$8.32	$332.80	6.5	$12.48	$81.12	$413.92	18%	$339.41
Dart	36	$9.28	$334.08	0	$13.92	0	$334.08	19%	$270.60
Franks	40	$9.48	$379.20	5.5	$14.22	$78.21	$457.41	19%	$370.50
Norris	40	$10.94	$437.60	7	$16.41	$114.87	$552.47	21%	$436.45

Total Regular Hours For All Workers __196__
(Add column 1)
Total Overtime Hours For All Workers __24__
(Add column 4)
Total Regular Pay For All Workers _$1,816.48_
(Add column 3)

Total Net Pay For All Workers _$1,741.02_
(Add column 9)
Total Overtime Pay For All Workers _$336.60_
(Add column 6)

Page 120

1. No. The total cost of the meal is $12.12. He has only $12.00.

2. $17.53

3. Tax and tip = $1.45. (The cost of her meal was $5.93, and $7.38 − $5.93 = $1.45.)

Page 121–122

1. a. $0.17
   b. $0.16
   c. $0.19

2. a. $0.48
   b. $0.49
   c. $0.47

3. a. $0.19
   b. $0.20
   c. $0.21

4. Home Foods: $14.35 for 15 pounds of grapes

5. Marcy's: $12.43 for 12 quarts of skim milk

Page 124

1. $r = d \div t = 182 \div 14 = 13$ mph

2. $d = rt = 59 \times 7 = 413$ miles

3. $t = d \div r = 3{,}255 \div 620 = 5.25$ hours, $15\frac{1}{4}$ hours, or 5 hours 15 minutes

4. $d = rt = 65 \times 9 = 585$ miles

5. Pam drove for $13 - 2 = 11$ hours
   $r = d \div t = 572 \div 11 = 52$ mph

6. $t = d \div r = 275 \div 55 = 5$ hours
   8:00 A.M. + 5 hours = 1:00 P.M.

7. Erik wants to drive for 5 hours.
   $r = d \div t = 265 \div 5 = 53$ mph

Page 125

1. $215.25

2. 175

3. $14.37 (14.368 to the nearest cent)

Page 127

1. 0.417 year

2. 3.083 years

3. 4.583 years

4. $225 ($i = \$2{,}500 \times 3\% \times 3$)

5. $330 ($i = \$1{,}500 \times 11\% \times 2$)

6. $71.72 ($i = \$750 \times 4.25\% \times 2.25$)

7. $408.99 (interest = $375 \times 3.75\% \times 2.417$)
   (total = $375.00 + $33.99)

8. $767.81 (interest = $650 \times 14.5\% \times 1.25$)
   total = $650 + $117.81)

Pages 128–129

1. 0.0060417 ($0.0725 \div 12$)

2. $540.73 ($89,500 \times 0.0060417$)

3. $89,430.18 ($89,500 + $540.73 − $610.55)

4.

Balance to Start Month	Interest for the Month	Balance to End the Month	Monthly Payment	New Balance
48,500.00	**272.81**	48,772.81	429.18	48,343.63
48,343.63	271.92	48,615.56	**429.18**	48,186.38
48,186.38	**271.05**	48,457.43	429.18	48,028.25
48,028.25	270.16	48,298.41	429.18	47,869.23
47,869.23	269.26	**48,138.49**	429.18	47,709.31
47,709.31	**268.36**	47,977.67	429.18	47,548.49
47,548.49	267.46	**47,815.95**	429.18	47,386.77
47,386.77	**266.55**	47,653.32	429.18	**47,224.14**
47,224.14	**265.64**	47,489.78	429.18	47,060.60
47,060.60	264.72	47,325.32	**429.18**	**46,896.14**
46,896.14	**263.79**	47,159.93	429.18	46,730.75
46,730.75	262.86	**46,993.61**	429.18	46,564.43

5. $3,214.58

6. $1,935.57 (48,500 – $46,564.43)

7. staying the same

8. decreasing

9. decreasing

10. $54,950.28

11. $91,426.83

12. $137,862.62

Pages 131–132

1. $P = 4 \times 4.1875 = 16.75 \approx 16.8$ inches
$A = (4.1875)(4.1875) \approx 17.5352 = 17.5$ square inches

2. $P = 2(8.9 + 3.2) = 24.2$ meters
$A = (8.9)(3.2) = 28.48 \approx 28.5$ square meters

3. $C = (2)(3.14)(5) = 31.4$ feet
$A = (3.14)(5)(5) = 78.5$ square feet

4. $P = 3.6 + 4.8 + 6.0 = 14.4$ inches
$A = \frac{1}{2}(6)(2.88) = 8.64 \approx 8.6$ square inches

5. $V = (2.0)(3.5)(4.5) \approx 31.5$ cubic inches

6. $V = (4.72)(4.72)(4.72) \approx 105.15$ cubic feet

7. $V = (3.14)(0.85)(0.85)(4.27) \approx 9.69$ cubic yards

8. $V = (3.14)(1.21)(1.21)(1.28) \approx 5.88$ cubic inches

9. I = 56.25 square units, II = 324 square units,
III = 380.25 square units

10. I = 14.0625 square units, II = 25 square units,
III = 39.0625 square units

11. 20.3 cubic inches $(2 \times 1.5) \times (2 \times 1.5) \times 10.5 - (3.14)(1.5)^2 \times 10.5$

Pages 133–134

1. $8 = \sqrt{64}$

2. $13 = \sqrt{169}$

3. $2.4 = \sqrt{5.76}$

4. 15; 20; 30; 3.16; 4.47

5. 4.8; 6.48; 3.95; 0.86; 0.3

6. 145; 66.49; 95

7. 5; 13.6; 91.65

8. 22.02; 7.91; 41.71

Page 136

1. [5] [×] [5] [M+]; 25.

2. [8] [×] [8] [M+]; 64.

3. [MR]; 89

4. [√] ; 9.4339811

5. 12.2 feet [7] [×] [7] [M+] [1] [0] [×] [1] [0] [M+] [MR] [√]

6. 13.9 feet [5] [×] [5] [M+] [1] [3] [×] [1] [3] [M+] [MR] [√]

7. 44 yards [3] [4] [×] [3] [4] [M+] [2] [8] [×] [2] [8] [M+] [MR] [√]

8. 7.5 miles [6] [·] [5] [×] [6] [·] [5] [M+] [3] [·] [8] [×] [3] [·] [8] [M+] [MR] [√]

Pages 137–140, Calculator Skills Review

1.

		Record All Charges or Credits That Affect Your Account						
Number	Date	Description of Transaction	Payment/ Debit (–)	✓ T	Fee (If Any) (–)	Deposit/ Credit (+)	Balance $ 641	90
309	7/6	Hillcrest Apartments	$ 275 00		$	$	366	90
310	7/7	Value Food Store	19 85				347	05
311	7/10	Northern Power Co.	107 33				239	72
312								
313	7/14	—VOID— Hair Palace	12 00				227	72
	7/16	Deposit Paycheck				342 61	570	33
314	7/18	2D Variety Store	27 93				542	40

REMEMBER TO RECORD AUTOMATIC PAYMENTS/DEPOSITS ON DATE AUTHORIZED

2.

		Wholesale Bathroom Supplies			
	Item #	Description	Quantity	Cost/Per	Total Amount
1.	A 241	Bath mat	14	$18.45	$258.30
2.	B 647	Bath towel	29	$10.98	$318.42
3.	D 832	Wall mirror	7	$73.49	$514.43
4.	F 301	Shower curtain	16	$19.99	$319.84
5.	P 970	Window curtain	8	$48.75	$390.00
				Total Purchase:	$1,800.99

3. Donnie's at $19.14

4. **a.** 22 trips; **b.** 192 bricks

5. $177

6. $6.52 \times \$12.50 = \$81.50$

7. $135.21

8. $2,100

9. $7.21

10. $5.52

11. 26%

12. $8,833

13. $2,415

14. **c.** $\$20 - (\$1.49 \times 3.5) - (\$1.19 \times 4.5)$

15. $9.43

16. 91 feet

17. **a.** 3.0 cubic feet
b. 22.5 gallons

Pages 158–160

1. 13,780

2. 667

3. $14.06

4. 5,772

5. 41

6. 216

7. The display shows 4; the calculator adds: $3 + 4 = 7$; the calculator multiplies: $2 \times 7 = 14$.

8. The displays are: 23; 23; 23; 23; 1; 24; 552; 2; 276.

9. The calculator shows: 1.5; 1.5; 2.26; 3.76; 2.14; 5.9; 5.9; 5; 1.18

10. $(23 + 48) \times 15.3$; 1086.3

11. $(5 \times 5.2) - (3.8 + 4.2 + 7.5)$; 10.5

12. $(7.1 \times 7.1) - (6.4 \times 6.4)$; 9.45

13. $(35 - (15.1 + 12.8)) \times 7$; 49.7

Page 161

1. 478.13

2. 2

3. 65.78

4. 2

5. 29.2

6. 18.67

Pages 162–163

1. 1.491
2. 2.189
3. 0.638
4. 4.519
5. 5.000
6. 1.343
7. 3.902
8. 0.006
9. 1.654
10. 8.000
11. 0.703
12. 6.328
13. 7.897
14. 0.417
15. 1.283
16. 0.099
17. 0.020
18. 0.010

Pages 164–166

1. 1.5707963

2. 94.24778

3. 0.1591549

4. 706.85835

5. **a.** $3\frac{1}{7}$ is greater
b. $3\frac{10}{71} = 3.1408451$
$\pi = 3.1415927$
$3\frac{1}{7} = 3.1428571$
$3\frac{10}{71}$ is closer to π; the difference is 0.0007476

6. $10! = 3,628,800$
$8! = 40,320$
$3! = 6$

7. 4

8. 9

9. 5

10. Answers may vary. Possible answer: The factorial in the denominator cancels out.

11. 16; 0.16; 16.81; 64; 144

12. 125; 1,296; 8; 512; 6,561

13. 11.56; 2,015.1121; 84.64; 314.432; 2,687.3856

14. 4,096; 97.336; 2,401; 27; 133.6336

Using Mental Math

Multiplying by Powers of Ten

The numbers 10, 100, 1,000, 10,000, and so on are called **powers of ten.** (They are so-named because $10 = 10^1$, $100 = 10^2$, $1,000 = 10^3$, $10,000 = 10^4$, and so on.) To multiply any number by a power of ten, count the zeros in the power of ten and write that number of zeros to the right of the other factor.

EXAMPLES

$$\begin{array}{r} 21 \\ \times\ \ 1,000 \leftarrow \text{3 zeros} \\ \hline 21,000 \leftarrow \text{product} \end{array} \qquad \begin{array}{r} 1,525 \\ \times\ \ \ \ \ 100 \leftarrow \text{2 zeros} \\ \hline 152,500 \leftarrow \text{product} \end{array}$$

To multiply a decimal number by a power of ten, count the zeros in the power of ten. In the other factor, move the decimal point *to the right* that number of places.

EXAMPLES $13.52 \times 1,000 = 13520.$

3 zeros   3 places

$0.05238 \times 100 = 05.238$

2 zeros 2 places

To multiply two factors with zeros at the right, count the zeros in both factors. Ignoring all the zeros, multiply the two numbers. Then write that number of zeros to the right of the product.

EXAMPLES

$$\begin{array}{r} 130 \\ \times\ \ \ \ 200 \\ \hline 26,000 \end{array} \quad \text{3 zeros} \qquad \begin{array}{r} 1500 \\ \times\ \ \ \ \ 4000 \\ \hline 6,000,000 \end{array} \quad \text{5 zeros}$$

13×2 3 zeros 15×4 5 zeros

Dividing with Zeros at the Right

To divide by a power of ten, count the zeros in the power of ten. In the numerator (the dividend), move the decimal point *to the left* that number of places.

EXAMPLES $\dfrac{575.2}{100} = 5.752$ $\dfrac{1.32}{10,000} = 0.000132$

2 zeros 2 places   4 zeros   4 places

To divide two numbers with zeros at the right, cancel the same number of zeros in the numerator as in the denominator.

EXAMPLES $\dfrac{600}{30} = \dfrac{60\cancel{0}}{3\cancel{0}} = \dfrac{60}{3} = 20$

$\dfrac{7000}{400} = \dfrac{70\cancel{00}}{4\cancel{00}} = \dfrac{70}{4} = 17.5$

Using Estimation

Front-End Estimation

In **front-end estimation,** you round each number to its left-most digit. Then you can use mental math to calculate with the rounded values.

ADDITION EXAMPLE

Find the rounded values by rounding each number to its left-most digit. Then add the rounded values.

Exact Problem		Rounded Values
687	→	700
4,435	→	4,000
+ 393	→	+ 400
		5,100 ← Estimate

SUBTRACTION EXAMPLE

Exact Problem		Rounded Values
483,478	→	500,000
− 217,675	→	− 200,000
		300,000 ← Estimate

MULTIPLICATION EXAMPLE

Exact Problem		Rounded Values
385	→	400
× 57	→	× 60
		24,000 ← Estimate

DIVISION EXAMPLE

Instead of rounding to the left-most digit, replace the numbers with approximate values that make the new problem divide evenly.

Exact Problem	Approximate Values	Reduced	Estimate
$\dfrac{529,420}{710}$ →	$\dfrac{490,000}{700}$ →	$\dfrac{490,0\cancel{0}\cancel{0}}{7\cancel{0}\cancel{0}}$ →	700

Formulas and Measurements

AREA		VOLUME		PERIMETER	
Rectangle	$A = lw$	Rectangular Solid	$V = lwh$	Square	$P = 4s$
Square	$A = s^2$	Cube	$V = s^3$	Rectangle	$P = 2(l + w)$
Circle	$A = \pi r^2$	Cylinder	$V = \pi r^2 h$	Triangle	$P = a + b + c$
Triangle	$A = \frac{1}{2}bh$			Circle (circumference)	$C = \pi d$

MEASUREMENTS

Celsius	Temperature	Fahrenheit
0°C	water freezes	32°F
100°C	water boils	212°F
37°C	normal body temperature	98.6°F

Time

60 seconds (sec) = 1 minute (min)
60 min = 1 hour (hr)
12 months (mo) = 1 year (yr)

24 hr = 1 day
7 days = 1 week (wk)
52 wk = 1 yr

Metric Weight

1,000 milligrams (mg) = 1 gram (g)
1,000 g = 1 kilogram (kg)

Customary

16 ounces (oz) = 1 pound (lb)
2,000 lb = 1 ton (T)

Length

1,000 millimeters (mm) = 1 meter (m)
100 centimeters (cm) = 1 m
1,000 m = 1 kilometer (km)

12 inches (in.) = 1 foot (ft)
3 ft = 1 yard (yd)
36 in. = 1 yd
5,280 ft = 1 mile (mi)
1,760 yd = 1 mi

Capacity

1,000 milliliters (mL) = 1 liter (L)
1 kiloliter (kL) = 1,000 L

3 teaspoons (tsp) = 1 tablespoon (tbsp)
1 fluid ounce (fl oz) = 2 tbsp
8 fl oz = 1 cup (c)
2 c = 1 pint (pt)
2 pt = 1 quart (qt)
4 qt = 1 gallon (gal)

ADVANCED CALCULATOR USE

This part of *Calculator Power* explains and explores some special calculator keys. See if these keys are on your calculator. If not, perhaps someone in your class has a calculator with these keys.

Two of the special keys that may be familiar are the left and right parentheses keys $($ and $)$ (also called the open parenthesis and close parenthesis keys).

Two other keys are the square root key $\sqrt{}$ and the reciprocal key $1/n$. The last four keys are π, $!$, x^2, and y^x.

Using Parenthesis Keys ⦗(⦘ and ⦗)⦘

In the last chapter of *Calculator Power* you explored parentheses in arithmetic expressions. To find the value of an arithmetic expression, you started by performing the operations inside the parentheses.

__EXAMPLE 1__ For the arithmetic expression 72 (53 − 18), first you find the value inside the parentheses.

	Press Keys	Display Reads
STEP 1 Find the difference 53 − 18.	⦗5⦘ ⦗3⦘	5 3 .
	⦗−⦘	5 3 .
	⦗1⦘ ⦗8⦘	1 8 .
STEP 2 Multiply that difference by 72.	⦗×⦘	3 5 .
	⦗7⦘ ⦗2⦘	7 2 .
	⦗=⦘	2 5 2 0 .

If your calculator has the parenthesis keys ⦗(⦘ and ⦗)⦘, you can find the value of the arithmetic expression by entering numbers and symbols from left to right.

	Press Keys	Display Reads
Enter each number or symbol. Be sure to enter a × symbol between the 72 and the ⦗(⦘.	⦗7⦘ ⦗2⦘	7 2 .
	⦗×⦘	7 2 .
	⦗(⦘	( ) 7 2 .
	⦗5⦘ ⦗3⦘	( ) 5 3 .
	⦗−⦘	( ) 5 3 .
	⦗1⦘ ⦗8⦘	( ) 1 8 .
	⦗)⦘	3 5 .
	⦗=⦘	2 5 2 0 .

Use a calculator that has the parenthesis keys ⦗(⦘ and ⦗)⦘ to find the value of each arithmetic expression.

1. $212 (147 - 82) =$

2. $(78 - 49) 23 =$

3. $(\$12.17 + \$32.19 + \$54.06) \div 7 =$

4. $78 (17 + 18 + 19 + 20) =$

5. $(39 + 41 + 43) \div 3 =$

6. $(12 + 15) \times (13 - 5) =$

Parenthesis Keys and the Display

When you use parenthesis keys, you can keep track of what the calculator is doing by looking at the value in the display.

EXAMPLE 2 Explain what is shown in the display when you find the value of this arithmetic expression: $(18 + 12) \times (19 - 11)$

Press Keys	Display Reads	Explanation
(	( )	The display shows parentheses.
1 8	( ) 18.	The display shows 18.
+	( ) 18.	No change.
1 2	( ) 12.	The display shows 12.
)	30.	The calculator adds $18 + 12 = 30$.
×	30.	No change.
(	( ) 30.	The display shows parentheses.
1 9	( ) 19.	The display shows 19.
−	( ) 19.	No change
1 1	( ) 11.	The display shows 11.
)	8.	The calculator subtracts $19 - 11 = 8$.
=	240.	The calculator multiplies $30 \times 8 = 240$.

For each of the following arithmetic expressions, write what the display shows as you enter each number or symbol key. Then explain what the calculator did to get the number that it displays. The first one is started for you.

7. $2 \times (3 + 4) =$

Press Keys	Display Reads	Explanation
2	2.	The display shows 2.
×	2.	No change.
(	( ) 2.	The display shows parentheses.
3	( ) 3.	The display shows 3.
+	( ) 3.	No change.
4	( ) 4.	_____
)	7.	_____
=	14.	_____

8. $23 \times (23 + 1) \div 2 =$

9. $\dfrac{1.5 + 2.26 + 2.14}{5} =$

Using Parenthesis Keys with Expressions

The next example shows how you can use the parenthesis keys to solve an expression.

EXAMPLE 3 $15 - (3.5 + 8.375)$

To solve this problem, subtract the sum of 3.5 and 8.375 from 15. Start with 15. Then find the sum of 3.5 and 8.375 and subtract that sum from 15.

		Press Keys	Display Reads
STEP 1	Start with 15.	1 5	15.
STEP 2	Press the keys as shown to subtract the sum of 3.5 and 8.375 from 15.	−	15.
		(	( ) 15.
		3 · 5	( ) 3.5
		+	( ) 3.5
		8 · 3 7 5	( ) 8.375
		)	11.875
		=	3.125

ANSWER: 3.125

....

Write an arithmetic expression for each number problem. Then use a calculator with parenthesis keys to find the value of the expression. The first one is started for you.

10. Multiply the sum of 23 and 48 by 15.3.
    Arithmetic expression: $(23 + 48) \times 15.3$
    Value: _____

11. Subtract the sum of 3.8, 4.2, and 7.5 from the product of 5 and 5.2.
    Arithmetic expression: _____
    Value: _____

12. Subtract the product of 6.4 times itself from the product of 7.1 times itself.
    Arithmetic expression: _____
    Value: _____

13. Subtract the sum of 15.1 and 12.8 from 35, and multiply the result by 7.
    Arithmetic expression: _____
    Value: _____

Your Choice: Memory Keys or Parenthesis Keys

Some arithmetic expressions can be evaluated using either memory keys or parenthesis keys. If your calculator has both, then you can decide which keys to use.

EXAMPLE　Evaluate this arithmetic expression using both memory keys and parenthesis keys. Compare the two answers.
$15.2 \times (7.5 + 3.2)$

	Press Keys	Display Reads
STEP 1 Use memory keys to calculate 7.5×3.2. Store the result in memory.	7 . 5	7.5
	+	7.5
	3 . 2	3.2
STEP 2 Enter the value 15.2 and multiply it times the value stored in memory.	M+	M 10.7
	1 5 . 2	M 15.2
	×	M 15.2
	MR	M 10.7
	=	M 162.64
To use the parenthesis keys, enter the numbers and symbols from left to right.	1 5 . 2	15.2
	×	15.2
	(	15.2
	7 . 5	M 7.5
	+	M 7.5
	3 . 2	M 3.2
	)	M 10.7
	=	M 162.64

ANSWER: Both sets of keystrokes give you the same value, which is **162.64.**

Solve each expression below using either memory keys or parenthesis keys. Round answers to the nearest hundredth.

1. $13.7 \times (15.1 + 19.8) =$

2. $44 \div (13.7 + 8.3) =$

3. $(23.5 - 18.9) \times 14.3 =$

4. $\dfrac{25}{6.26 + 6.24} =$

5. $7.53 + (6.07 \times 3.57) =$

6. $100.4 - (12{,}037 \times 6.79) =$

Using the Reciprocal Key $\boxed{1/n}$

Here is a straightforward calculation.

Divide 53 into 75.

There are at least three ways to write an arithmetic expression for this calculation.

$53\overline{)75}$ $\dfrac{75}{53}$ $\dfrac{1}{53} \times 75$

For the third choice, $\frac{1}{53} \times 75$, notice that "divide 53 into 75" has been rewritten as "multiply $\frac{1}{53}$ by 75." The fraction $\frac{1}{53}$ is called the **reciprocal** of 53. If your calculator has a reciprocal key $\boxed{1/n}$, you can use it to solve both division and multiplication problems.

EXAMPLE 1   Use the reciprocal key to divide 53 into 75. Round your answer to the nearest hundredth.

	Press Keys	Display Reads
STEP 1 Enter 53 and then press $\boxed{1/n}$.	$\boxed{5}$ $\boxed{3}$ $\boxed{1/n}$	0.0188679
STEP 2 Enter $\boxed{\times}$ and $\boxed{7}$ $\boxed{5}$.	$\boxed{\times}$ $\boxed{7}$ $\boxed{5}$	75.
STEP 3 Press the $\boxed{=}$ key.	$\boxed{=}$	1.4150943

ANSWER: 1.42

Use the reciprocal key to perform each calculation. Round your answers to the nearest thousandth.

1. Divide 171 into 255.

2. Find $\frac{1}{132} \times 289$.

3. What is 235 divided into 150?

4. Find $\frac{1}{233} \times 1,053$.

5. Divide 17.3 into 86.5.

6. Find $\frac{1}{15.35} \times 20.62$.

7. Divide the sum of 41.7 and 13.6 into 215.8.

8. What is the reciprocal of 175?

9. Divide 0.0052 into 0.0086.

10. What is the reciprocal of 0.125?

The following example shows how the reciprocal key can be used to evaluate a **complex fraction.** A complex fraction is a fraction in which the numerator or the denominator or both contain fractions (or decimals).

EXAMPLE 2 Use the memory keys and the reciprocal key to evaluate the complex fraction. Round your answer to the nearest hundredth.

$$\frac{432-173}{47.1+38.8}$$

	Press Keys	Display Reads
STEP 1 Evaluate the numerator (the top of the fraction) and place the value in memory.	4 3 2	432.
	−	432.
	1 7 3	173.
	M+ *	M 259.
STEP 2 Evaluate the denominator (the bottom of the fraction) and find its reciprocal.	4 7 . 1	M 47.1
	+	M 47.1
	3 0 . 8	M 30.8
	=	M 85.9
	1/n	M 0.0116414
STEP 3 Multiply by the value stored in memory.	×	M 0.0116414
	MR	M 259.
	=	M 3.0151339

ANSWER: 3.02

....

 If your calculator has memory keys and a reciprocal key, use them to evaluate each complex fraction. Round your answers to the nearest thousandth.

11. $\dfrac{127-23}{85+27+36} =$

12. $\dfrac{\left(12.8+3.5\right)\left(4.6-1.3\right)}{\left(1.4+1.1\right)\left(5.8-2.4\right)} =$

13. $\dfrac{15\left(8.7+9.2\right)}{17\left(3.1-1.1\right)} =$

14. $\dfrac{4+\frac{1}{5}}{10+\frac{1}{15}} =$

Find the value of each arithmetic expression. Round your answer to the nearest thousandth.

15. $\frac{1}{2}+\frac{1}{3}+\frac{1}{4}+\frac{1}{5} =$

16. $\dfrac{1}{10+\frac{1}{10}} =$

17. $\frac{1}{10}-\frac{1}{11}+\frac{1}{12}-\frac{1}{13}+\frac{1}{14}-\frac{1}{15} =$

18. $\dfrac{1}{100+\frac{1}{100}} =$

*On some calculators, you may need to press = before M+.

Using Four More Special Keys

Two keys that are usually found on scientific calculators are the pi key π and the factorial key $!$.

The Pi Key π

The value referred to as π is a fixed, constant number. It is the ratio, in any circle, of the circumference to the diameter. The value of that ratio, to the ten-millionths place, is 3.1415927.

 If your calculator has a π key, use it to perform each calculation.

1. $\frac{\pi}{2}$

2. What is $2 \times \pi \times r$ if $r = 15$?

3. $\frac{1}{2\pi}$

4. What is $\pi \times r \times r$ if $r = 15$?

5. The value of π is between the two mixed numbers $3\frac{1}{7}$ and $3\frac{10}{71}$.

   a. Which value is greater, $3\frac{1}{7}$ or $3\frac{10}{71}$?
   b. Which value is closer to π? By how much?

The Factorial Key $!$

How many ways can five persons stand in line for a photograph? Since anyone can stand at the left of the line, there are five choices. That leaves four persons, so there are four choices for the next position in the line. Continuing, you get

Choice of 5 persons		Choice of 4 persons		Choice of 3 persons		Choice of 2 persons		Choice of 1 person		
5	×	4	×	3	×	2	×	1	=	120 ways

An expression that is the product of all numbers from a given whole number down to the number 1 is called a **factorial.** The mathematical symbol for factorial is the exclamation point, !.

EXAMPLE 1

"4 factorial" = $4 \times 3 \times 2 \times 1 = 4 \boxed{!} = 120$

"7 factorial" = $7 \times 6 \times 5 \times 4 \times 3 \times 2 \times 1 = 7 \boxed{!} = 5,040$

"1 factorial" = $1 = 1 \boxed{!} = 1$

Use a calculator with a $\boxed{!}$ key to perform each calculation.

6. 10!, 8!, and 3!

7. $\frac{4!}{3!}$

8. $\frac{9!}{8!}$

9. $\frac{5!}{4!}$

10. Explain any patterns you see in problems 7 through 9.

The Keys $\boxed{x^2}$ and $\boxed{y^x}$

The $\boxed{x^2}$ key is called the **square key.** It gives you a quick way to find the square of any number.

Expression	Method 1	Method 2	Result
5^2	$\boxed{5}\ \boxed{\times}\ \boxed{5}\ \boxed{=}$	$\boxed{5}\ \boxed{x^2}$	25.
17^2	$\boxed{1}\ \boxed{7}\ \boxed{\times}\ \boxed{1}\ \boxed{7}\ \boxed{=}$	$\boxed{1}\ \boxed{7}\ \boxed{x^2}$	289.
$(1.36)^2$	$\boxed{1}\ \boxed{\cdot}\ \boxed{3}\ \boxed{6}\ \boxed{\times}\ \boxed{1}\ \boxed{\cdot}\ \boxed{3}\ \boxed{6}\ \boxed{=}$	$\boxed{1}\ \boxed{\cdot}\ \boxed{3}\ \boxed{6}\ \boxed{x^2}$	1.8496

The $\boxed{y^x}$ key is called the **power key.** Here is how it is used.

$$\underbrace{4 \times 4 \times 4 \times 4 \times 4 \times 4}_{6 \text{ factors}} = 4^6 \ \leftarrow \text{exponent}$$

base

The **exponent** 6 tells how many times the **base** 4 appears as a factor. The entire expression 4^6 is called a **power.**

Expression	In Words	Meaning	Press Keys	Value
5^3	five to the third power*	$5 \times 5 \times 5$	$\boxed{5}\ \boxed{y^x}\ \boxed{3}\ \boxed{=}$	125
2^4	two to the fourth power	$2 \times 2 \times 2 \times 2$	$\boxed{2}\ \boxed{y^x}\ \boxed{4}\ \boxed{=}$	16
$(1.3)^2$	1.3 to the second power**	1.3×1.3	$\boxed{1}\ \boxed{\cdot}\ \boxed{3}\ \boxed{y^x}\ \boxed{2}\ \boxed{=}$	1.69

*A number raised to the third power, such as 5^3, can also be read as "five cubed."
**A number raised to the second power, such as $(1.3)^2$, can also be read as "1.3 squared." To evaluate an expression with an exponent of 2, you can use the $\boxed{x^2}$ and press $\boxed{1}\ \boxed{\cdot}\ \boxed{3}\ \boxed{x^2}$.

Use your calculator to find the value of each problem below. For another way to solve these problems, read the Discovery below.

11. 4^2 $\qquad$ $(0.4)^2$ $\qquad$ $(4.1)^2$ $\qquad$ 8^2 $\qquad$ 12^2

12. 5^3 $\qquad$ 6^4 $\qquad$ 2^3 $\qquad$ 8^3 $\qquad$ 9^4

13. $(3.4)^2$ $\qquad$ $(6.7)^4$ $\qquad$ $(9.2)^2$ $\qquad$ $(6.8)^3$ $\qquad$ $(7.2)^4$

14. 8^4 $\qquad$ $(4.6)^3$ $\qquad$ 7^4 $\qquad$ 3^3 $\qquad$ $(3.4)^4$

Discovery: Most four-function calculators have a **multiplication constant** feature. This feature gives another way to find the value of a power. Instead of repeatedly pressing the two-key combination of the number and ⊠, press and re-press the equals key =.

The following examples will show two ways to find the value of a power.

Expression	Repeated Multiplication	Multiplication Constant
7^2	7 × 7 =	7 × =
7^3	7 × 7 × 7 =	7 × = =
7^4	7 × 7 × 7 × 7 =	7 × = = =

After entering 7 × ,

• Pressing = once gives the value of 7 to the second power.

• Pressing = twice gives the value of 7 to the third power.

• Pressing = three times gives the value of 7 to the fourth power.

Glossary

A

and   The word that represents the decimal point in a decimal number or amount of money. For example, 4.054 is read "four *and* fifty-four thousandths"; $12.43 is read "twelve dollars *and* forty-three cents."

arithmetic expressions   Expressions that involve numbers, parentheses, and the operations +, −, ×, and ÷. For example, (34 + 23) ÷ 3. Note: In the phrase "arithmetic expression" the word "arithmetic" is used as an adjective, so the accent is on the next-to-last syllable: a rith **met** ic.

C

calculate   Work with numbers

calculator   An electronic device that makes it easy to work with numbers

calculator letters   The symbols that look like letters when a calculator display is read upside down. The calculator letters are I (1), Z (2), E (3), h (4), S (5), L (7), b (9), and O (0).

clear all key   The key that clears any value stored in memory and any value displayed by the calculator

clear display key   The key that clears the value displayed by the calculator. It does not affect any value stored in memory.

clear memory key   The key that clears any value stored in memory. It does not affect any value displayed by the calculator.

common fraction   A number expressed as a ratio of two integers. For example, $\frac{3}{4}$, $\frac{-7}{10}$, and $\frac{115}{-8}$ are common fractions.

D

decimal   A number between zero and one expressed with a decimal point. 0.15 and 0.1119 are decimals.

decimal places   Places to the right of a decimal point. The decimal places are tenths, hundredths, thousandths, ten-thousandths, hundred-thousandths, millionths, ten-millionths, and so on.

difference   The answer to a subtraction problem. In the problem 14 − 9 = 5, the difference is 5.

digit keys   The keys for the digits 0, 1, 2, 3, 4, 5, 6, 7, 8, 9

display   The panel on a calculator that shows the current value calculated by the calculator

dividend   The number you are dividing into in a division problem. In the division problem $35 \div 7 = 5$ or $7\overline{)35}$, the dividend is 35.

divisor   The number you are dividing by in a division problem. In the division problem $35 \div 7 = 5$ or $7\overline{)35}$, the divisor is 7.

E

error symbol   An indication in a calculator that it has been asked to do something it cannot do. A calculator displays an error symbol if the number being calculated is too large to be shown in the display window (this is called an **overflow**) or if it has been asked to divide by zero.

estimating   Using rounded values to find an approximate answer to a problem

F

four-function calculator   A calculator that has keys for the four operations (or functions) of addition, subtraction, multiplication, and division. Most four-function calculators also have memory keys.

I

improper fraction   A fraction that is greater than or equal to 1. $\frac{15}{11}$ and $\frac{7}{7}$ are improper fractions.

K

keying error   The act of making incorrect keystrokes. For example, entering [8] [2] instead of [2] [8], entering [1] [0] [3] instead of [1] [·] [3], entering [7] [7] [4] instead of [7] [4], and so on

M

memory keys   Keys for storing and retrieving values entered by the user or calculated by the calculator

mental math   Performing a calculation in your mind, without writing it down or using a calculator

mixed decimal   A decimal number that combines an integer and a decimal. 3.055, 100.9, and –34.732 are mixed decimals.

mixed number   The sum of a whole number and a proper fraction. $3\frac{2}{7}$ and $113\frac{67}{70}$ are mixed numbers.

O

on/off key   The key that turns a calculator on or off

overflow error   An indication in a calculator that it has been asked to calculate a value too large to be shown in the display window

P

part   In the percent problem "25% of 200 is 50," the value 50 is the part.

percent   In the percent problem "25% of 200 is 50," the value 25% is the percent.

percent circle   A device that can be used to help remember which operation to perform to solve various kinds of percent problems

percent key   The key that shows the results of a calculation as a percent. [5] [6] [×] [2] [5] [%] displays 14.

perfect square   A number that is the product of some number times itself. For example, 64, 100, and 225 are perfect squares because $64 = 8^2$, $100 = 10^2$, and $225 = 15^2$.

place value   Places to the left or right of a decimal point. To the left of the decimal point the place values are ones, tens, hundreds, thousands, and so on. To the right of the decimal point the place values are tenths, hundredths, thousandths, ten-thousandths, hundred-thousandths, millionths, ten-millionths, and so on.

placeholding zero   Using the digit zero to help digits line up in a calculation. The addition problem at the right uses placeholding zeros.

```
  24.58          24.580
 112.046        112.046
+ 78.3         + 78.300
```

problem-solving steps
1. Read the problem and find the key information.
2. Estimate an answer.
3. Choose the operation and set up a calculation.
4. Perform and check the calculation(s).
5. Reread the problem and verify your answer.

product   The answer to a multiplication problem. In the problem $3 \times 6 = 18$, the product is 18.

proper fraction   A fraction in which the numerator (top number) is smaller than the denominator (bottom number). $\frac{2}{3}$, $\frac{177}{178}$, and $\frac{5}{358}$ are proper fractions.

Q

quotient   The answer to a division problem. For example, in the division problem $35 \div 7 = 5$ or $7\overline{)35}$, the quotient is 5.

R

reciprocal   A fraction with 1 in the numerator and a particular number in the denominator. The reciprocal of 3 is $\frac{1}{3}$; the reciprocal of 12 is $\frac{1}{12}$; the reciprocal of $\frac{3}{4}$ is $1 \div \frac{3}{4}$ or $1 \times \frac{4}{3}$ or $\frac{4}{3}$.

remainder   In a division problem, the part of the quotient that is not a whole number. For example, in the division problem $41 \div 7 = 5$ r 6 or $7\overline{)41}$, the remainder is 6.

repeating decimal   A decimal number that goes on forever using a pattern of digits. 0.23232323 . . . is a repeating decimal; 1.875 is not a repeating decimal (it is a **terminating** decimal).

rounded number   A number with zeros to the right of a given place value. 700 is the rounded number for 742.

S

square root key   The key that finds the square root of a displayed number. $\boxed{3}$ $\boxed{6}$ $\boxed{\surd}$ displays 6.

sum   The answer to an addition problem. In the problem $4 + 9 = 13$, the sum is 13.

T

terminating decimal   A decimal number that has a limited number of digits. 1.875 is a terminating decimal; 1.333 . . . is not a terminating decimal (it is a **repeating** decimal).

terms   Numbers, products, and quotients in an arithmetic expression. In the expression $\frac{27 + 2 \times 7}{13} + 23 \times 5 - 16.4$, there are three terms: $\frac{27 + 2 \times 7}{13}$, 23×5, and 16.4.

truncate   To express a long decimal as a short one by "cutting off" the digits at some particular place. If the decimal number 5.88888 . . . is represented by 5.88, it has been truncated. (If it is represented by 5.89, it has been **rounded.**)

W

whole   In the percent problem "25% of 200 is 50," the value 200 is the whole.

Index

A

Arithmetic expressions 91, 93, 95
Average 125

C

Calculators
    battery-powered 8
    four-function 7, 8
    rounding 47
    solar-powered 8
    truncating 50
Consumer problems 29, 65, 85, 120, 121, 125, 126, 128

D

Decimals
    mixed decimal 45
    reading decimals 45
    repeating decimals 49
    rounding decimals 47
    terminating decimals 49
    truncated decimals 50

E

Estimation 20, 25, 27, 32, 34, 35, 54, 57, 60, 62, 63, 155

F

Formulas 156
    distance, rate, time 123
    measurement 130
    simple interest 126
Fractions
    improper fractions 66, 70
    mixed numbers 66, 68
    proper fractions 66
    changing to decimals 66

K

Keying errors 17, 20
Keys
    clear key 9
    decimal point key 10
    digit keys 7
    equals key 15
    factorial key 164
    function keys
        add 15, 19
        divide 33
        multiply 31
        subtract 21
    memory keys 89
    parenthesis keys 158
    percent key 75, 77, 79, 81
    pi key 164
    power key 165
    reciprocal key 162
    square key 165
Key words 52

M

Math intuition 25, 35
Mental math 154
Multiplication constant 166
Multistep problems 37, 97, 100

P

Patterns 51
Percents
    percent circle 73, 75, 77, 79
    percent increase 81, 100
    percent decrease 81, 100
Perfect squares 133
Pythagorean theorem 135

S

Square roots 133

W

Word problem-solving steps 23
Workplace problems 39, 114, 115, 116, 117, 118